To geologists everywhere,
especially the ones whose hands are dirty...

••••••••••••••••••••••••••••••••••••

and with special thanks to Gustav Mahler
for help with the subtitles
and for sending along the Resurrection Symphony.

••••••• BOOKS 1 AND 2 •• | •••••••••• BOOK 3 •••••••••• | •••••••••••••••••• BOOK 4 •••••••••••••••••••••• | ••••••••••••••

CONTENTS

PROLOGUE

BOOK ONE
SECRETS OF THE STRATA

BOOK TWO
BIRTH OF THE EARTH-MOON SYSTEM
4550–4500 MY AGO

BOOK THREE
FIRST DAYS ON THE SURFACE OF A NEW WORLD
4500–3500 MY AGO

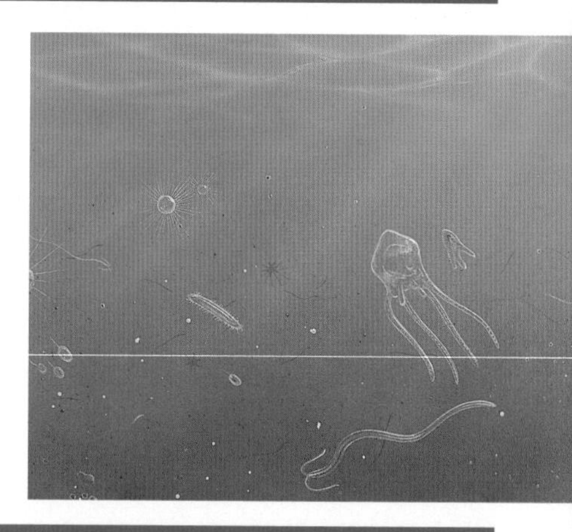

BOOK FOUR
AN UNFAMILIAR EARTH AND A MIDLIFE CRISIS
3500–1000 MY AGO

THE
HISTORY
OF
EARTH

A view from the red-hot surface of the forming moon, 4500 million years ago. The sky is dominated by a red-hot Earth, recently remolded and melted by the giant planetesimal impact that produced the moon. The ring of debris is aggregating into countless moonlets, which in turn are eventually accreted by the moon.

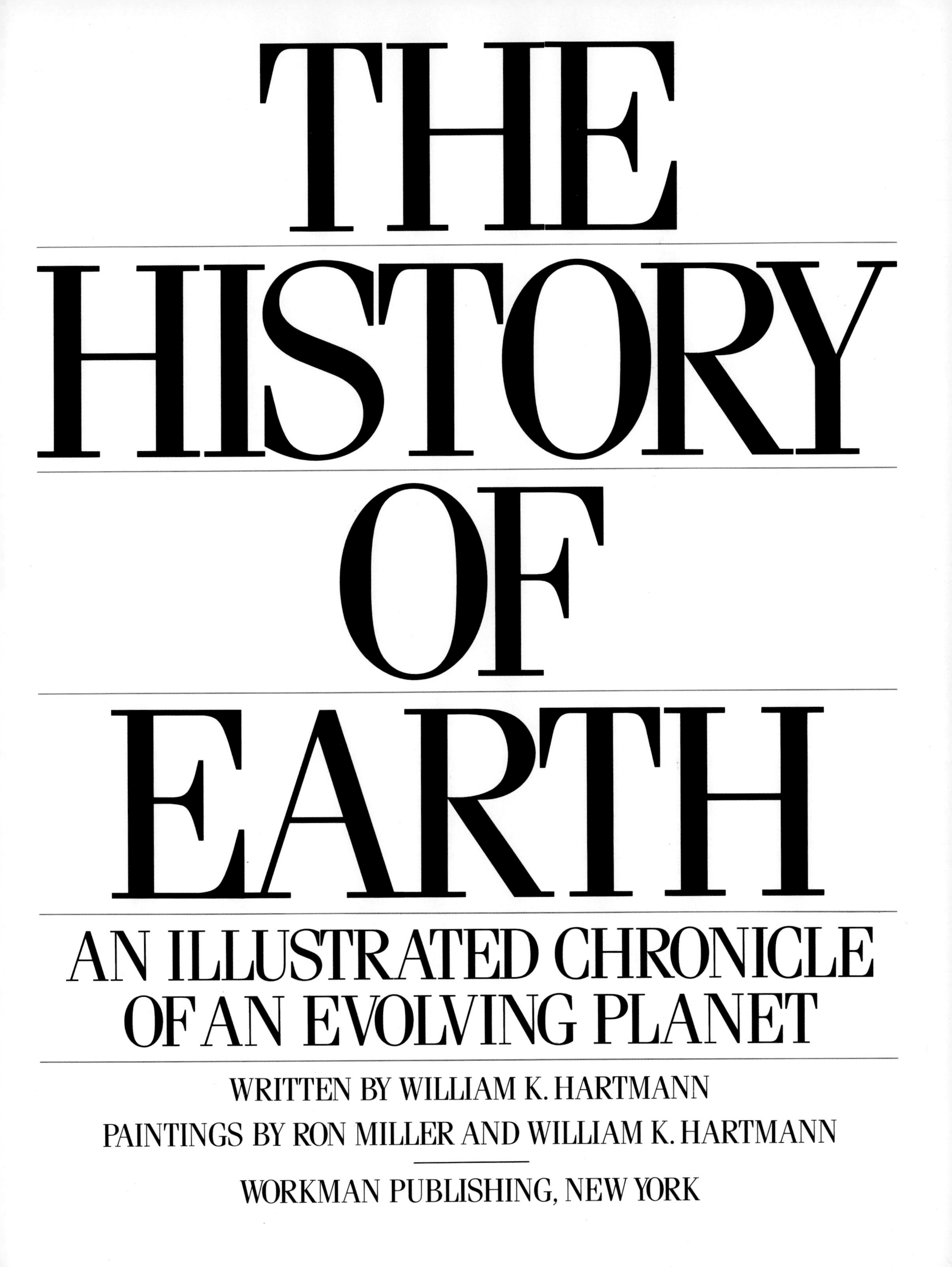

THE HISTORY OF EARTH

AN ILLUSTRATED CHRONICLE OF AN EVOLVING PLANET

WRITTEN BY WILLIAM K. HARTMANN

PAINTINGS BY RON MILLER AND WILLIAM K. HARTMANN

WORKMAN PUBLISHING, NEW YORK

Acknowledgments

We wish to thank Gayle Hartmann and Judith Miller for assistance and critiques during preparation of the manuscript and artwork. Special thanks also to Alan Hildebrand, Mike Drake, Lisa McFarlane, and Steffi Engel for discussions of the Cretaceous-Tertiary boundary events, the formation of Earth's mantle, and early organic chemistry. Of course, unfortunately, responsibility for all factual blunders and gross misunderstandings, if any, reside with us (though each of us reserves the right to blame the other). Finally, thanks to Sally Kovalchick, Julie Hansen, Teresa Quirk, Nancy Inglis, Tom Starace and the staff at Workman Publishing for assistance in preparation of the book.

W.K.H. and R.M.

Published simultaneously in Canada by Thomas Allen & Son Ltd.

Library of Congress Cataloging in Publication Data

Hartmann, William K.
The History of Earth: an illustrated chronicle of an evolving planet / by William K. Hartmann and Ron Miller

1. Earth—Origin. 2. Earth—Origin—Pictorial works. 3. Geology. 4. Geology—Pictorial works.
5. Cosmology. 6. Cosmology—Pictorial works.
I. Miller, Ron II. Title
QB632.H37 1991 525—dc20 91-50387 CIP

ISBN 1-56305-122-2
ISBN 0-89480-756-0 (pbk.)

Workman Publishing Company, Inc.
708 Broadway
New York, NY 10003

First printing November 1991

10 9 8 7 6 5 4 3 2 1

Manufactured in Hong Kong

BOOK FIVE

EARTH MATURES

1000–65 MY AGO

BOOK SIX

THE MOST UNUSUAL ONE PERCENT OF GEOLOGIC TIME

65 MY AGO TO THE PRESENT

BOOK SEVEN

BEYOND OUR TIME

THE PRESENT TO 6000 MY FROM NOW

There is not only more than one history of the world,
one for each of us who studies it;
there is more than one for each of us,
there are as many as we want or need,
as many as our heads and wanting hearts can make.

—John Crowley,
Aegypt, *1987*

Inventing Geology

No transformation in men's attitude to Nature—in their "common sense"—has been more profound than the change in perspective brought about by the discovery of the past. Rather than take this discovery for granted, it is almost preferable to exaggerate its significance.

—*Stephen Toulmin and June Goodfield, quoted in* Language of the Earth, *by F. Rhodes and R. Stone, 1981*

Geology is relatively new. The idea that Earth is a beautiful, fragile, evolving ball is also new. By "new," I mean that these ideas have evolved in only the last couple of centuries. In this book we will trace the evolution of our 4,500,000,000-year-old Earth itself, but first a few words about the evolution of our *ideas* of Earth.

As we approach the year 2000, we are in a unique era apropos planetary perceptions. We have gone beyond the age of superstition, but also beyond the age of pure pragmatic exploitation. We are past the age of Earth's so-called endless frontier. We may even be passing the we-they age of environmentalist vs. engineer. With the exception of some national leaders who seem to be national followers, we seem to be approaching a consensus that our attitude toward Earth will have to change. We are approaching an age of synthesis, an age in which we can combine the best aspects of many different perceptions. We can see our blue Earth not only as provider of raw materials but also as mother and nourisher of life. At the same time we

Hartmann photo

Humanity began to manipulate its environment as soon as culture emerged. This 4000-year-old site was a ceremonial center used, at least in part, to calibrate the calendar each year at the summer solstice. Stone alignments mark the horizon position of sunrise on the solstice date. Stonehenge, England.

can see Earth as our sister in beauty, creator of rose-streaked sunsets, leaves, and the translucent curls of ocean waves. We can appreciate Earth not only as our home but also as the fairest of many planets and as a source of scientific information about planetary origins as well as climatic and biological evolution. More than any other generation, ours is able to see all these Earths simultaneously—and indeed *must* do so to survive in an age of environmental crisis.

Just as a guidebook helps us appreciate the geography and heritage of a new city, a guidebook to Earth's evolution helps us appreciate our whole world and respond to its unique beauty more deeply. We Americans, having no Stonehenges in our meadows and no Gothic-age cathedrals on our street corners, tend to forget that modern "enlightened" civilization, with its appreciation of scientific curiosity, is a frosting of attitudes and discoveries layered atop a dark cake of medieval superstition and apathy. Curiosity about the world has not been universal in other times and places. For example, Charles M. Doughty (1843–1926), the British explorer of Arabia some years before Lawrence, encountered some mysterious ruins in the desert and asked his Arab guides about them. "These are just works remaining from the creation of the world," the Arabs replied. "What profit is there to inquire about them?"

Here was a response which in its lack of curiosity sums up the indifference of many peoples to the features of the world around us. It is an indifference characteristic of many cultures that did not experience Greco-European history—the Greek discovery of science and then its resurgence during the Renaissance only a few centuries ago. We should be glad to live in an age when the fantastic tale of our planet is being discovered. But the old indifference lurks in America today, when many citizens have become dulled to the pleasures of understanding the world around them, and when college science enrollments have dropped to around the levels of the 1960s.

EARTH, SCIENCE, AND PSEUDOSCIENCE

All right. Given that Western culture developed a scientific way of reconstructing the history of Earth, why should we waste any further time talking about it? If we're going to read about geology at all, why not plunge right into the good stuff about Earth's formation, volcanoes, and dinosaurs?

Patience. Stay for a moment to talk about history, science, and civilization. Our story will be richer if we see how geology got started, and how it produced a whole set of ideas that are almost subliminally buried in today's daily papers and TV programs. Anyone can recite some facts about Earth and its history; fewer understand *why* we believe these facts. Armed with that information, you have not just fact but understanding! The *why* involves the history of science itself.

The veneer of modern science in our society is seen to be very thin if we look at it in historical perspective. The intertwined fields of geology and biology have been under attack since their beginnings by fundamentalists of one sort or another—those who formed their conceptions of Earth not by looking at the planet itself but by looking at old texts, whether they be Greek, Moslem, Christian, Marxist, or Hindu, or writings from still other traditions.

For several centuries, fundamentalists have maintained the myth of a past golden age of received wisdom (the era of the Garden of Eden, ancient Egypt, Aristotle's Greece, or whatever), when divine messengers or Great Authorities recorded fundamental truths; the world, so the myth usually goes, has declined ever since. Assuming that truth is revealed best by their favorite old book, fundamentalists don't just shun modern scientific evidence, they attack it. The scary thing is that this still goes on in our generation, whether from neo-Nazis burning non-Aryan books and avant-garde art, from Mideastern fanatics threatening modern novelists, from Marxist bureaucrats banning open scientific discussion of ideas that might threaten their economic theories, or from American "creation scientists" bringing lawsuits to require the teaching of their own favorite Christian scriptural interpretations in public schools. The self-styled creation scientists, or creationists, present their attack on science as if it were something new, but the history of geology reveals that they are only the newest recruits in a long battle against the open-minded study of our planet.

The famous Scopes trial of 1925, which tried to end the teaching of evolution in Tennessee schools, was by no means the last such attempt. The attack goes on today. Some state school boards have proposed the teaching of fundamentalist versions of geology and biology as an alternative to science. Several lawsuits that would have restricted the teaching of modern geology and biology were narrowly turned back in the courts during the eighties in the United States. The veneer of science in our culture is thin.

Religious extremists who attack Darwinian evolution in the courts and media might be aghast to discover that they are cousins of Stalin, who also tried to replace Darwinian evolution with his own favorite theory, which he thought matched Marxist dogma better. A professor named Trofim Denisovich Lysenko (1898–1976), who was a favorite of Stalin's, advocated the idea that if you work hard enough to develop a certain trait, your children can inherit that trait—independent of traditional Darwinian natural selection or genetics. Stalin thought this matched his ideas of the emergence of the "new Soviet man." For this ideological reason, he forced Lysenkoism on the universities, setting Soviet biology back by many years. Only after Stalin's death did Soviet biology begin to mesh with scientific developments in other countries.

If the creationists succeed in America, more than just the general comprehension of biology and life will be eroded; the significance of fossils and the chronology of strata will be dismissed, and our understanding of our planet itself will decline, just at the point when we need it most.

If this picture of science and civilization under attack surprises some readers, that in itself is a good reason to begin with a historical overview of Earth science. A great naturalist and early geologist of the last century, Louis Agassiz (1807–1873), put it more succinctly and with a touch of humor:

> Every great scientific truth goes through three stages. First, people say it conflicts with the Bible. Next, they say it has been discovered before. Lastly, they say they always believed it.

SORTING WHEAT FROM CHAFF—THE STRENGTH OF REAL SCIENCE

As Agassiz said, we accept seemingly "obvious facts" about Earth as if everyone had known them forever. But as we review the history of geology, you will witness teams of combatants struggling back and forth, shaping the very ideas that now seem obvious. Following these tournaments of the mind is as exciting as watching the latest bowl game on TV and leaves you with something more worthwhile: insight into the world around us.

In scientific bowl games, the scores are not the results of lucky breaks or referees' decisions; they accrue from skillful assembly of observations. But in science the game is not about winning and losing, any more than someone wins when a jigsaw puzzle is completed. In science, finding the right evidence to put on the table is the name of the game. You need no lawyers to advocate your case; the evidence does that for you. And all the players can celebrate when the bits of evidence suddenly click into place, like pieces of a puzzle.

But how do we know which are the right pieces? Bookstore shelves often carry as much crackpot pseudoscience as valid science. One way to distinguish one from the other is to look at the degree to which the author is in contact with other leading researchers. For example, scientific papers submitted to scientific journals are generally sent out for review by others who work in the field; they must receive approval by at least two referees before being published. Papers are usually discussed and debated by other researchers after publication. Also unlike pseudoscience, the assertions in scientific papers are backed up by references to other papers in leading journals so that skeptical readers can go back to the earlier work to judge for themselves. An author backed up by other leading researchers may not necessarily be right, but at least you have some assurance that he/she is not raving alone, off in left field.

According to a recent study (see Hamilton, 1991), about 60 percent of research papers in the physical, chemical, biological, and geological sciences are listed within five years as references in other papers by other scientists in the leading technical journals worldwide. Virtually zero percent of publications from the pseudoscientific fields, such as "creation science" and astrology, are referenced in such journals. In other words, scientists are constantly talking to each other, debating the merits of one another's claims, and weeding out the claims that don't withstand scrutiny. The maundering pseudoscience you see at the supermarket checkout stands is not even part of that dialogue. This explains why, throughout history, the claims and predictions of geologists and biologists have generally held up better than the conflicting claims and predictions of their pseudoscientific, fundamentalist critics. How many times have fundamentalists predicted the end of the world, and how many times have they been right? How many geothermal energy sources have been found through their approach?

TRUTH, BEAUTY, AND FOSSILS

Two parallel streams of thought led to our current ways of looking at Earth. One involved ways of getting at the truth, and the other, curiously enough, involved ways of appreciating beauty. One of the last instances in which these two streams were both major concerns of a single person was around 1500, when Leonardo da Vinci was both an artist and a naturalist. He kept voluminous notebooks of observations and sketches of rivers, rocks, clouds, waves, skeletons, and other phenomena. As he developed the roots of modern science, he simultaneously worked on his art without considering it a different type of activity. Sketching and painting, for Leonardo, were part of the scientific method—a way to understand nature.

Among Leonardo's many interests was geology. One day he was walking in the mountains of Italy and came upon some fossil seashells. He looked at these more inquiringly than most people. He noted them in his journal as "testimony of things

500 years ago. Leonardo da Vinci recorded some of the first modern geological observations. Visiting the northern Italian Alps, he recorded well-preserved fossils of marine creatures, from which he correctly deduced that ancient sea-floor strata had been uplifted to mountaintop positions.

produced in the salt waters and found again in the high mountains, far from the seas of today." Leonardo wanted to know how sea creatures' remains could have gotten so far above sea level. Some thinkers of his day assumed that they had been driven inland by storms of the biblical Deluge. But Leonardo noted that the more delicate shells would have been broken by such violence, as can be seen on any modern beach. Thus he concluded that these fossils lay in soils where the creatures had originally lived.

> I perceive that the surface of Earth was from old entirely filled and covered over in its plains by the salt waters, and that the mountains, the bones of Earth, [were made by uplifting the old ocean floors to their present altitudes]. . . . Subsequently, the incessant rains . . . by repeated washing, have stripped bare part of the lofty summits . . . . [Soil from the] summits . . . has . . . already descended to their bases, and has raised the beds of the seas . . . , and in some parts has driven away the seas from there, over a great distance.

These ideas were basically correct. Just as you can arch a deck of cards by pushing on the ends of the deck, massive deformations of Earth bend the originally level crust and raise part of it into high plateaus. Mountains are the eroded bones of such high plateaus. Many present-day mountains are formed of strata that once lay on the sea floor, studded with shells of sea creatures, and were later thrust upward to high elevation, only to be eroded into rocky crags.

Leonardo's ideas of a continually changing Earth were a far cry from conventional wisdom of his day. People thought of Earth's surface as an unchanging platform for the march of history. Even a century later, Shakespeare wrote that "all the world's a stage." Life, Shakespeare said, is a poor player who frets his hour upon the stage and then is heard no more. Life seemed dynamic and transient, but Earth's stage seemed as static as the floorboards of the Globe Theatre.

Eventually the idea of a static Earth was overturned. By the time of the American and French revolutions, naturalists had established the radical idea that the Earth was not just a passive backdrop, but a scene of constant change. Excitement spread through scientific circles as geologists realized that by careful observation they could find clues to past events and decipher Earth's story.

Ideas of how to seek truth and beauty continued to evolve. When observant wanderers such as Leonardo walked in the mountains for recreation, they not only invented the methodology of seeking answers directly from nature but also helped change the way people respond to nature. The change was revolutionary. Forays into the wilderness to observe nature culminated in a radical idea: even the wildest parts of nature are beautiful.

This was a new perception of the planet, a change in viewpoint that we rarely think about today. Such a change may seem amazing, because most people assume that the way we respond to Earth's beauty is a fixed law of the human psyche, no less fundamental than the law of gravity. It is more realistic to say that our ideas of beauty have evolved along with culture. In the Middle Ages, and even at the peak of urbane rationalism in the 1700s, civilization was viewed as good; wild, dangerous nature was viewed as bad. Hence the old myths of the dark forest, full of witches and trolls. Nature was ominous and unpredictable. In Sergei Prokofiev's version of *Peter and the Wolf*, Grandfather warns Peter not to go out of the garden: "What if a wolf should come out of the forest? What then?"

It is natural that earlier generations thought nature to be dangerous. It was. The very mountain vistas that delight today's traveler were looked upon with dread. Mountains were abodes of brigands and unrest, spawning grounds of snowstorms and avalanches. They were dreadful obstacles to travel. The surprising thing to us is that mountains were thought to be not just dangerous but ugly. There were subtle reasons. In the 1600s, scholars thought that Earth started as a beauteous Garden of Eden and had since decayed. This ideology implied that mountain ranges were blemishes on an aging Earth, a conception that affected perceptions of our planet. Geological historian G. L. Davies quotes a traveler in 1693 who crossed the Alps and reported them to be "horrid, ghastly ruins." Similarly, art historian Kenneth Clark remarks that the Italian poet Petrarch, in the 1300s, was the first person in recorded history to have climbed a mountain for pleasure. In this connection, Clark calls Petrarch the first modern man.

Examining such historical attitudes, we reach the astonishing conclusion that the pleasure we derive from Earth is not an absolute human trait, but derives from our whole philosophic value system.

The old dread of nature began to change as naturalists began to tromp mountains and cliffsides in search of clues to Earth's history. By the era of Beethoven, around 1800 to 1820, young poets, writers, naturalists, and musicians crystallized this incredible revolution in the way we looked at beauty. It was the *romantic revolution*. Instead of celebrating the staid pleasures of city life, manicured lawns, prim behavior, repressed emotions, and nature under control, they celebrated the wild, the unpredictable. They emphasized the word *sublime*, a word that has been devalued by semantic inflation. Sublime, which might be used today to describe something no more exciting than a down pillow, meant to the romantics the manifestation of God's and Nature's most awesome effects. Storms, lightning, earthquakes, the ineffable colors of an aurora swirling overhead—these were sublime effects. The romantics put themselves in direct sensual contact with nature. They taught us that to *experience* Earth for recreation re-created the psyche.

Today, in our relation to Earth, we are schizophrenic about the romantic rebellion and the classical view it replaced. Much twentieth-century environmental writing, for example, portrays nature as good and civilization as bad. That view contains some truth (practically every statement contains some truth, if you look hard enough), but it is not very productive. What we will need in our future on Earth is the ability to synthesize and enjoy both approaches to nature. We will need to *choose*—to design—the kind of civilization we want so that we can enjoy life and sustain Earth at the same time. When you enjoy the beauty of a good photograph or the sensuality of music played from your latest CD, you are responding to romanticism. When you admire the engineering of the beautifully built camera or CD player, you are responding to classicism—the well-ordered approach to life and craftsmanship that makes technology possible. The two streams of thought may come together: When you take the camera and CD player to the mountains to photograph the sunset's glow on a snowy peak while you listen to music, you are responding to both romanticism and classicism. And if you muse on a fossil you find there, you are following in Leonardo's footsteps. If we combine all these strains of thought, we are prepared to respond to the science, beauty, and mystery of Earth. It is the planet with the most fantastic and complex biography among all the known planets. Here, with more unexpected twists and turns than in a science-fiction tale, is the story of a world that started as lifeless as an asteroid, had its offspring moon ripped from its own body, has not always had a twenty-four-hour day, has spent most of its years with no plants or beasts on its desolate continents, underwent a drastic change of atmospheric composition in middle age, and only in the last fraction of a percent of its history has been inhabited by upright land creatures who immodestly call themselves intelligent.

BOOK ONE

Secrets of the Strata

1. The First Geologists

> After three years of almost daily warfare with Dr. Hutton on the subject of his theory, I began to view his fundamental principles with less and less repugnance.
>
> —*Sir James Hall,* Transactions of the Royal Society of Edinburgh, *1812*

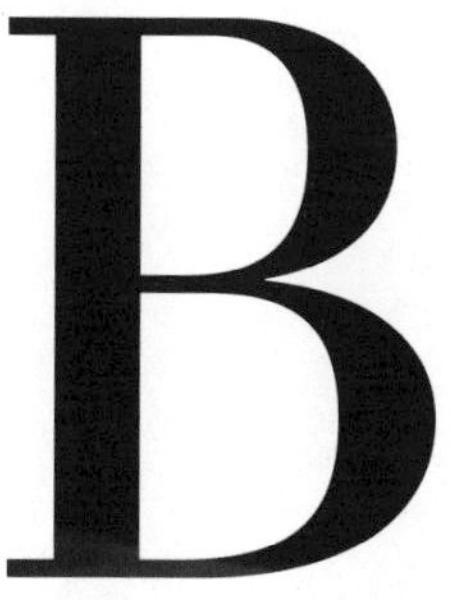

By the late 1700s a number of people were looking at Earth as a giant book of clues. One pioneer of this new idea was a German naturalist, Abraham Werner (1750–1817), who studied how layers of different types of rock formed successive strata. In the 1780s he correctly hypothesized that each stratum represented certain rock-forming conditions during Earth's history. At last, theories about Earth being derived directly from observation. Not just the art of geology but the *science* of geology was being born.

NEPTUNISM VS. PLUTONISM—ORIGINS OF MODERN GEOLOGY

Werner advanced one of the first grand geologic theories. Today we recognize the difference between *igneous rocks*—rocks formed by solidification of molten material—and *sedimentary rocks*—rocks composed of fine particles that settled out of water and were deposited in layers. But in Werner's day, the origins of these rock types were uncertain. Although Werner was correct in concluding that different layers revealed an unfolding story of different formative conditions, he was wrong in his guess as to how the rocks formed.

Geothermal areas are regions where the subsurface temperatures are hotter than normal, due to magma activity at depth. One of the most famous is Yellowstone National Park, Wyoming, shown here.

Werner thought all rocks formed as sediments deposited in oceans. He pictured the primordial Earth as covered by a vast worldwide ocean in which a succession of different rocks was laid down by precipitation, like chemicals separating out of a solution, and by sedimentation, or settling of fine particles. Generally, he thought that igneous rocks had formed first, followed by sedimentary rocks. Slowly the ocean must have retreated, exposing massive sedimentary layers of strata, such as limestones and sandstones. On top of these, erosion carved valleys and deposited gravel. The most recent rocks of all, in Werner's theory, were the volcanic lavas, such as were actually being observed to form in Italy. Werner's students liked to point out, in support of the theory, that his ancient global ocean more or less agreed with the idea of the biblical flood. Because of the key role of the ocean, Werner's theory was given the name Neptunism, after the Roman god of the sea.

In the same decade as Werner worked, the 1780s, the Scottish naturalist James Hutton (1726–1797) came up with a different theory from his observations of the Scottish Highlands. According to Hutton, many rock types had not been deposited as sediments in the sea but rather had crystallized from a molten state. Other rock types were created by heating and remelting of the first types. Because heat played a role in it, Hutton's theory gained the title Plutonism, after the Roman god of the underworld.

Modern geology was born in the following few decades as adherents of Neptunism and Plutonism argued back and forth. By the 1820s the battle was all over. Studies of Italian volcanoes indicated that their fresh lavas were identical to old lavas in France that Neptunists had mistakenly called seafloor deposits. The fact that these ancient French rocks must have formed from molten material was a victory for Plutonism.

Most geology textbooks present Plutonism as the clear winner and cite examples of the clear-eyed Huttonians arguing from natural examples, while the Neanderthal Wernerians, blinded by their belief in the flood, refused to see the light. Today we can be a bit more charitable, recognizing that humans seem to have a need to polarize the truth in order to make it easier to talk about, and that the truth, as is usually the case, was more subtle. Hutton had correctly recognized that igneous rocks form directly from crystallization and solidification of molten material. The most important such rocks are a dark brown or gray lava called basalt and a lighter-colored crystalline rock called granite. Hutton also pointed out that some of these rocks have been altered by heat and pressure, and often distorted so that their minerals form wavy bands. These altered rocks are called *metamorphic rocks*. But Hutton tended to overplay rock formation from molten magmas and to underplay the role of rock formation from the cementing together of particles precipitated out of water. Werner was correct in pointing out that many sedimentary rocks, such as limestone and sandstone, were deposited in water without any significant role for plutonistic heating. As is often the case in a violent argument, each side was seeing only a part of the truth.

Instead of trying to decide if one side was right and one side was wrong, we should celebrate an important outcome of the debate: the three major types of rocks—igneous, sedimentary, and metamorphic—were firmly recognized, and are still being taught to every geology student today.

MAPPING THE GEOLOGIC PERIODS

As early geologists began to recognize the three major rock types, they made an even more profound discovery. Different groups of fossils could be recognized in different beds of sedimentary rock. One stratum might have a certain group of fossil fish, but a layer above it might have different fossil fish. This meant that fossils could be used to identify specific beds. A particular fossil group in a gray limestone stratum, for example, allowed a geologist to identify and follow that layer from one hillside to another, even if it was exposed in different parts of the landscape miles apart.

Geologists recognized that strata are deposited one atop another, so lower layers are generally older than upper layers.* Careful charting of vertical sequences in flat-lying beds gave geologists a reliable guide to the time sequence of various fossils.

*This rule is safest in flat-lying strata. In occasional circumstances, geologic forces may have contorted the strata—for instance, making huge S-shaped folds—so that an older unit in a local exposed area, such as a road cut, appears above a younger unit. Field geologists must be wary at all times!

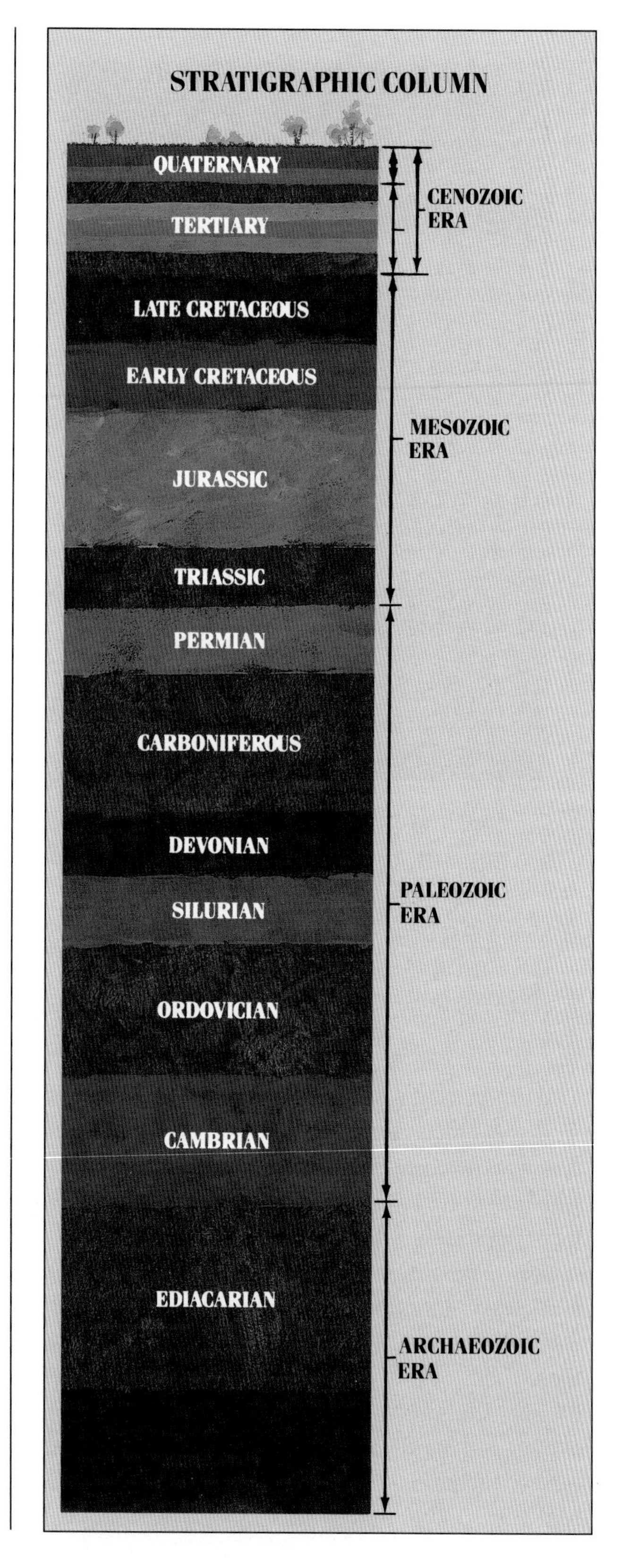

Geologists soon discovered that certain classes of fossils were characteristic only of specific time periods. They were called index fossils. The time periods were named after areas where the index fossils were commonly found. For instance, English geologists in Cambria and Devon named the Cambrian and Devonian Periods, while researchers in the Jura Mountains of Switzerland and France named the Jurassic Period. To take a simplified example, certain species of fish might commonly be found together in the Devonian strata, but not in strata from other periods. Wherever those fossil fish were found together, whether in a limestone in one region or in a sandstone in another region, the layer of host rock must have formed during the Devonian Period. The differences in rock from region to region might tell us that during the Devonian a sea occupied one region while the delta at a river's mouth occupied another.

New models of Earth's history came with the mapping of rock sequences in the early 1800s. It was clear, for instance, that the Cambrian layer was deeper and older than the Devonian layer. Assembling their data, geologists were gradually able to arrange the different suites of fossils, and their periods, in an age sequence called the *stratigraphic column*. Thus geologists knew the *relative* ages of different formations and events in Earth's history, although no one knew the *absolute* ages, as measured in years.

Periods were grouped in three broad eras: the Paleozoic, Mesozoic, and Cenozoic Eras, or the era of ancient life, middle life, and recent life, respectively. Many early geologists assumed that the oldest fossils dated from very early in the planet's history. More recent scientists, however, have learned that the Paleozoic, Mesozoic, and Cenozoic Eras mark only the last segment of geologic time. For most of Earth's history, there were no lifeforms advanced enough to leave many fossils. That long, pre-fossil era is called the Archaeozoic Era.

It is important to remember that despite all their progress in the recording of fossils, none of the first geologists knew how far back in time each period or era had begun, or for how many years it lasted.

CATASTROPHISM VS. UNIFORMITARIANISM

Even as the controversy between Neptunists and Plutonists died down, a new controversy arose. One group of early geologists—the *catastrophists*—believed that dramatic, violent, sudden changes produced Earth's major features. Another group—the *uniformitarians*—argued that nature's processes operated at a more or less uniform rate; Earth's features could be explained in terms of the slow processes we see operating today, but extending over vast eras of time.

The controversy is worth reviewing because it is still with us today. This statement will surprise some readers because all geology textbooks say that the uniformitarians emerged triumphant, and that catastrophism was demolished. Indeed, all geology students have the principle of uniformitarianism pounded into their heads. Usually this is reduced to a simple statement:

THE PRESENT IS THE KEY TO THE PAST.

This means that structures produced in ancient times can be explained in terms of processes observed today. Hutton, the Scottish geologist who was the father of Plutonism, is credited with being the father of uniformitarianism as well. In a 1788 paper, he phrased the principle thus: "The operations of nature are equable and steady."

One type of formation cited by the catastrophists in support of their ideas was the puzzling case (found in several locations) of a mountain ridge cut through by a river. The well-known Delaware Water Gap is a good example of such a structure in the United States. Catastrophists assumed that a mighty cataclysm must have cut the ridge to allow the water to break through. Or perhaps a cataclysmic upthrust raised the halves of the mountain on either side of the river. Uniformitarians correctly argued that the original level of the landscape was much higher, and that the humble process of stream erosion, grain by grain and pebble by pebble, removed all the soil, with the river always occupying the lowest spot. The river had established its course on the original surface of the ground, at higher elevation, and then slowly eroded its way down through the rocky layer now marking the ridge, no doubt forming waterfalls and rapids where it crossed the resistant layers, as the surrounding land level was worn away. The mountain ridge was left standing on either side because it was a more resistant rock type.

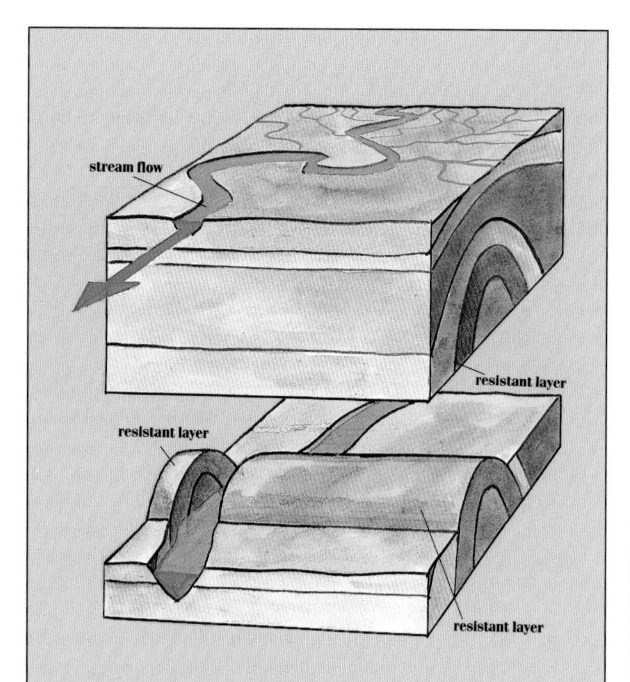

Development of a notch-like canyon such as the Delaware Water Gap, which puzzled early geologistis. A meandering stream cuts down through sediments (top). It may pond behind resistant rock layer but will eventually erode through, creating a notch (bottom). Such formations prove that some modern landscapes are results of erosion of enormous thicknesses of sediments that once covered the region.

The argument between catastrophists and uniformitarians continued as other geologists pointed out features that they thought came not from the slow workings of such familiar processes as stream erosion, but from abrupt events. By the 1820s, the Swiss-French geologist Georges Cuvier (1769–1832) was emphasizing the sensational discovery that some lines of fossil animals had become extinct. Ancient fossils included species never seen by humans! Cuvier wanted to know how this could be if "the operations of nature are . . . steady." Moreover, it seemed to him that some species had died out with dramatic suddenness, only to be replaced by other species.

Today we know that some of these cases of seemingly sudden replacement are geological illusions. For example, if fossil-bearing strata are deposited on the sea floor and then the sea retreats, the land may be exposed to erosion for, say, 100 million years. During those 100 million years, the youngest layers of sea-floor strata and their fossils may be completely removed. Then suppose the sea advances again and new fossil-bearing strata are deposited on top of the old. Now, a geologist looking at the layers at this site will see a tremendous break in the fossil record—a break consisting of the part that was eroded. But the gap was caused only by a long period of erosion, not by a catastrophic change.

But Cuvier was not to be dismissed easily. He quoted another bit of evidence for catastrophism, citing in an 1825 paper the recent discoveries of perfectly preserved extinct mammoths frozen in the arctic tundra. Because of the specimens' excellent state of preservation, he felt they must have been living healthily in benign environments, only to be suddenly and perfectly frozen. He noted that they must have been frozen as soon as they died; otherwise rotting would have begun.

Today these observations no longer seem like arguments for mysterious catastrophes. We can easily

A good example of the gradual process of erosion is shown here. As the tree grew, erosion removed soil around the roots, exposing the upper root system. Bryce Canyon National Monument, Utah.

imagine mammoths grazing in the arctic tundra in early fall, only to be trapped in mud and frozen in a sudden winter storm. A catastrophe for the mammoth, perhaps, but not for the whole continent. Nonetheless, the case of the frozen mammoths is interesting because it still crops up in crackpot literature as evidence of supposed baffling geological mysteries or as evidence for ancient global disasters.

Still another example cited by catastrophists involved the fossils Leonardo da Vinci found on the mountaintop and the implied uplift of onetime sea-floor strata by thousands of feet from sea-floor beds to elevated ranges. For example, a colleague of Cuvier pointed to sea-floor fossils at 6000 feet in the Alps, similar to other fossils found in the plains around Paris. Wearing away by uniformitarian erosion could not explain elevated seashells. What agent could cause such uplift? Catastrophists cited disastrous earthquakes. They were much impressed by the 1755 quake that had destroyed much of Lisbon, Portugal, and killed more than 10,000 people. Based on this, they assumed that even greater ancient quakes could have raised whole mountain ranges. Today these ideas have been abandoned; we have evidence for long, slow uplift processes that were unknown to geologists of the early 1800s.

Writing in 1827, Cuvier gave a nod to uniformitarianism, but said it was not enough:

> . . . it has long been considered possible to explain the more ancient revolutions on [Earth's] surface by means of . . . still existing causes. . . . But we shall presently see that unfortunately this is not the case. . . . The thread of operations is here broken. . . , and none of the agents that [nature] now employs were sufficient for the production of her ancient works.*

One of the major factors in this historical argument is that most catastrophists thought Earth was young. Catastrophist theory, which started out as scientific research, got tangled up with biblical interpretations by leading theologians, who believed that Earth and its landscapes were created in a few abrupt steps (the six days of creation), and who estimated that Earth was only a few thousand years old. One of the most famous estimates of the Earth's age was that of English Archbishop James Ussher (1581–1656), who through detailed studies of scripture had been able to "prove" that Earth was created on October 26, 4004 B.C. at 9:00 A.M.! The idea that Earth was only 6000 years old underlay most of the early geological discussions. Naturally, for people who thought Earth was only 6000 years old, the mighty Alps and its craggy valleys could be explained better by fast-acting catastrophes than by erosion acting at immeasurably slow rates.

In 1835 the famous naturalist Charles Darwin

*Hallam (1983) cites Cuvier at greater length and discusses the whole early period in more detail. Also fascinating and informative is the detailed historical discussion by Davies (1969). These and other references are found in the references at the end of the book.

1 My ago. Erosion by flowing water has been an important sculptor of Earth since the planet formed. The most famous example of its work in North America is the Grand Canyon. The Colorado Plateau was broadly and gently uplifted from about 100 to 50 My ago, and the river has cut rapidly downward through the flat-lying sediments, exposing an unparalleled cross section of the stratified rocks.At the bottom of the canyon are Precambrian rocks more than 500 My old.

(1809–1882) observed a Chilean earthquake that raised the coastline a few feet for a stretch of 100 miles. On a plain 250 feet above the beach he noticed fossils similar to those at the shore. Back in England in 1838, Darwin presented a paper on this phenomenon. He reasoned that the long, slow cumulative effects of such earthquakes could explain the raised beach 250 feet above. One such earthquake every century, for example, could have raised the 250-foot cliff in 10,000 years, and could raise a 25,000-foot Himalayan peak in a million years. Uniformitarian principles worked if Earth was old enough.

For the time being, no one knew Earth's age, but uniformitarians successfully argued that Earth must be *much* older than the conventional estimate of a few thousand years.

The final victory of uniformitarianism is usually attributed to the work of the Scottish-English geologist Charles Lyell (1797–1875). Hutton's writings were wordy and opaque, but Lyell wrote clearly and enthusiastically, and became a tireless champion of Hutton's ideas. Lyell traveled widely, studying Italian volcanoes and other geological features and summarizing his views in the multivolume *Principles of Geology,* which first appeared in 1830 and went through eleven editions, the last appearing in 1875. It was Lyell who established the uniformitarian principle not only as a theory of Earth's history but also as a method. The theory was that most changes had happened slowly and gradually by processes observable in the present, and the method was to study those present-day processes in order to explain the past.

THE NEW CATASTROPHISM

Even though catastrophism is usually considered a dead subject, the argument between catastrophism and uniformitarianism still echoes around us. Fascinatingly, it rears its head at two opposite poles of society: among the subcultures of the religious fundamentalists and also among a new wave of geoscientists. The former are usually rehashing two-century-old arguments, and the latter are discussing the latest geologic observations.

The reason for the fascination of fundamentalists with catastrophism seems to be due not so much to their clear-eyed appraisal of modern evidence (what they would have us believe) as to the half-forgotten origins of catastrophism alongside biblical interpretation in the 1790s. Early fundamentalists and catastrophists found mutual support in Earth's supposedly short age and sudden divine acts of creation and punishment (the Garden of Eden, the creation of Adam, the destruction of the world by the flood, the destruction of Sodom and Gomorrah). Biblical scholars, who accepted the idea of Earth's development in abrupt steps (the six days of creation), found more comfort in catastrophism than in the slow, mindless workings of uniformitarian processes. The ascendancy of the ideas of long-time-scale evolution and uniformitarianism was much to the horror of fundamentalists as well as catastrophists.

This is why fundamentalists of the 1990s look back longingly, and sometimes unconsciously, to the catastrophists' arguments of 1800. I have heard late-night radio evangelists exhuming the old arguments once again, condemning modern geological dating of ancient strata and brandishing articles about Cuvier's frozen mammoths as if they were a profound new challenge to modern science. And a graduate-level "creation research" school, licensed by California to award degrees in geology, biology, and other fields, hands out Master of Science degrees for thesis research that, in the words of the prestigious journal *Science* (February 15, 1991), "assumes fossils are the remains of drowned critters left behind by Noah's Ark on a planet that is less than 10,000 years old."

In that climate, modern ideas of catastrophic changes in Earth's past can easily be misunderstood. During most of the twentieth century, geology textbooks have extolled Hutton/Darwin/Lyell-style uniformitarianism and have stated that catastrophes were not needed to explain geological findings. The 1980s changed everything. Close observations of the stratigraphic record revealed that radical changes in species really did occur, over periods of not more than a few million years. This seems long by human standards but is only a wink of the cosmic eye.

What caused the radical changes? As we will discuss in detail in Chapter 13, scientists now realize that Earth and other planets have been hit occasionally by wandering asteroids and comets—rocky and icy chunks of interplanetary debris. Some of these random impacts were probably large enough to affect Earth's climate and wipe out species during geologically short periods. Thus, unknown to earlier geologists, who taught their students that "the present is the key to the past," a few events in Earth's history really were catastrophic. The new ideas are sometimes called "the new catastrophism," or "neo-catastrophism."

This is not a revival of nineteenth-century catastrophism because it does not support a short time scale for Earth's history. Earth's major continent-shaping processes still happen at the old, slow, uniformitarian rates. The sudden changes are simply superimposed on the background of slow change. The new catastrophism gives no comfort to the fundamentalists, but instead makes a profound link between the history of Earth and the history of the whole solar system. Like the Apollo astronauts' photos of our globe floating in space, the new catastrophism catapults Earth out of its nineteenth-century isolation and places it where it should be, in a rich cosmic environment.

2. How Old Is a Rock, and How Old Is the Planet?

The poor world is almost six thousand years old.

—*William Shakespeare, ca. 1600*

he study of Earth's history was firmly launched by the mid-1800s. By then, geologists had powerful tools—the classification of major rock types, the use of fossils to establish the stratigraphic column that showed the sequence of different rock layers, and the use of uniformitarian principles to interpret how landscapes formed.

A century elapsed before discovery of the second great interpretative tool of geology: a technique for using radioactivity to measure the absolute ages of rocks in years. This technique allowed geologists to date not only the relative ages of the Cambrian, Devonian, and other fossil-defined periods but also their absolute ages and the age of Earth itself.

The ages of rocks and planets involve enormous amounts of time—far more than the early catastrophists or even the uniformitarians dreamed. Evolution of mountain ranges takes millions of years. Evolution of life involved thousands of millions of years. Earth itself turned out to be 4,500,000,000 years old.*

Mathematicians and scientists express this unwieldy age of 4,500,000,000 as 4.5×10^9 years, but this usage intimidates some readers. In America, a very common usage has been to say the Earth is 4.5 billion years old. There's a problem with this solution, however: In British usage the word *billion* means 1,000,000,000,000, or a million million. In American usage it is only 1,000,000,000, or a thousand million. For this reason, international scientists a few years ago agreed on a new standard system of prefixes. *Giga-* is the prefix for the American billion, so that 4,500,000,000 years is called 4.5 Gigayears, or 4.5 Gy. *Mega-* is the prefix for million, and one million years is referred to as 1 My. The American billion is 1000 My, and Earth's age can thus be written as 4500 My.

A million years is a good unit of geological time—a sort of geological moment. In fact, if you represent the history of Earth by a 24-hour day, each million years would come out as about a third of a minute. Therefore, in this book, we will express most dates in units of millions of years, abbreviated My. This will avoid the Anglo-American confusion about the meaning of *billion* and will also prevent us from having to switch units in mid-book, as we move from a discussion of early time, billions of years ago, to the more detailed geologic record of recent time, only millions of years ago.

*Ages greater than 1,000,000,000 years first began to be discussed in 1905, soon after the discovery of radioactivity. Rough estimates for the age of Earth at around 4,000,000,000 were published by the American geologist T. C. Chamberlin in 1920 and the American astronomer H. N. Russell in 1921, and a direct measurement of 4,500,000,000 was first published by the American geologist C. C. Patterson in 1953. (See the historical reviews by Brush and by Badash in the references.)

WHAT ATOMS TELL US

How can a geologist pick up two rocks, take them into the laboratory, and come out later to state that "this rock is four hundred My old, but that rock is only thirty My old?"

A rock is an assemblage of mineral crystals bonded together. Each crystal is a chemical compound made up of certain atoms. Quartz crystals, for instance, are the compound silicon dioxide, or SiO_2, meaning that each molecule in quartz's crystal structure has one atom of silicon and two of oxygen. Many other crystals are more complicated, containing many atoms of silicon, oxygen, iron, potassium, magnesium, and so on. The method of dating rocks is based on clues hidden in the atoms that make up each of the crystals of a rock.

I have spoken of the dating of rocks as if there were a single technique. Actually, there are several

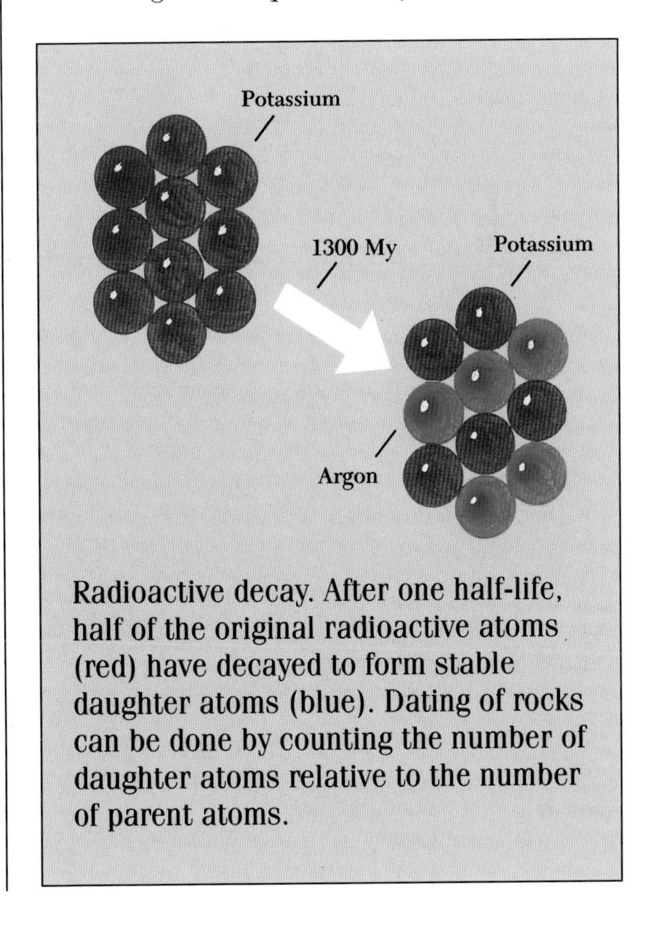

Radioactive decay. After one half-life, half of the original radioactive atoms (red) have decayed to form stable daughter atoms (blue). Dating of rocks can be done by counting the number of daughter atoms relative to the number of parent atoms.

As water evaporates from landlocked lakes, calcium carbonate deposits are left, forming grotesque structures known as tufa. Mono Lake, California.

related techniques, each based on different kinds of atoms. However, all the techniques are based on the fact that certain types of unstable atoms, called radioactive atoms, suddenly and spontaneously convert themselves to other types by giving off subatomic particles. This happens at a certain rate, which can be measured in a lab. The essence of the dating technique is to count up how many atoms have converted from the so-called parent type to the so-called daughter type, and then figure out how long it took for this to happen.

For simplicity, consider the potassium-argon technique, one of the most popular. This is based on the slow conversion of radioactive potassium into the gas argon. It works best on rocks that initially contain abundant potassium and in which the gaseous daughter argon atoms are trapped in the rock during long periods of time.

Potassium and all other elements have different forms, called isotopes. Only certain isotopes are radioactive—that is to say, unstable. Other isotopes are stable.

A single atom, of course, is the smallest possible unit of any given element, and each atom must be one isotope or another of that element. Inside each atom is its nucleus. The nucleus has some protons and some neutrons. All isotopes of a given element have the same number of protons in their nuclei. The number of protons in the nucleus defines the element. For example, any atom with nineteen protons is an atom of potassium. Atoms have roughly the same number of neutrons as protons. However, the number of neutrons can vary slightly. For example, the most common stable form of potassium has twenty neutrons, but there are other forms with nineteen, twenty-one, twenty-two, and even more neutrons. Each different form is called an isotope. About 93 percent of all potassium atoms have twenty neutrons and are stable, but a tiny fraction of potassium atoms in nature are the isotope with twenty-one neutrons, and it is unstable. Given enough time, these atoms spontaneously change themselves one by one into the element argon. It is this radioactive form of potassium that is useful for dating rocks.

Geochemists use shorthand to refer to the different isotopes. Each isotope is designated by the letter abbreviation for its element (K in the case of potassium, Ar for argon) and a number that gives essentially the total number of protons and neutrons. Thus the radioactive form of potassium is designated as K^{40}.

Half the K^{40} atoms in any crystal will convert themselves into Ar^{40} in 1300 My (1,300,000,000 years, or 1.3 billion years to Americans). This is an important figure. It is called the half-life for that isotope. Each radioactive isotope has a different half-life. Carbon 14, which is used in dating archaeological deposits, has a half-life of only 5730 years. The important point is that after one half-life, half the radioactive parent atoms will have converted into daughter atoms.

Now the basic technique for dating becomes clear. Consider molten rock material, which is called magma. As long as it is molten, any gas atoms, such as argon, can escape from it. As long as the magma is molten, it is essentially argon-free. Besides, argon is so chemically nonreactive that it will not combine in any crystals that form. Now the magma cools and starts to crystallize. The crystals at first contain no argon. But consider a certain type of mineral crystal that contains potassium. Imagine a microscopic bit of crystal that starts out with 1000 K^{40} atoms and no Ar^{40} atoms. Then, after 1300 My, this crystal would have 500 K^{40} atoms and 500 Ar^{40} atoms. After another half-life, half of those 500 K^{40} atoms would have also changed to argon. Thus 2600 My after the crystal formed, there would be 250 K^{40} atoms and 750 Ar^{40} atoms. And so on.

All the geologist has to do to measure the age of a rock is get the rock to a very sophisticated geochemistry laboratory and measure the number of Ar^{40} atoms in a crystal, relative to the number of K^{40} atoms.

In practice, as always, there are complications. Suppose for example, despite our best hopes, some argon gas atoms had been in the spaces between atoms of the crystal as it formed. How would we know which Ar^{40} atoms were truly daughter atoms from the conversion of K^{40}, and which were there initially? Fortunately, there are geochemical techniques for overcoming such problems, involving the measurement of the proportions of the other argon isotopes. Through techniques of this type, very refined estimates of the ages of different kinds of rock can be obtained with errors of only a few percent or less.

Because these ages are derived by using different types of radioactive isotopes, they are usually called *radioisotopic ages*. It is important to notice exactly what such an age represents. In the case of a potassium-argon age, we are clearly measuring the time since the rock began trapping the gas argon. Generally, this means the time since the rock solidified.*

Most other types of radioisotopic ages, based on elements other than potassium, also measure the time since the rock solidified. Notice that if our rock sample remelted, perhaps due to volcanic activity, the argon would escape. Then, as the molten material cooled and resolidified into a new rock, the "clock" would be reset and the trapping of radioactively created argon would recommence. Thus the radioisotopic age always measures the time since the most recent solidification event.†

THE GEOLOGIC TIME SCALE

Once reliable radioisotopic ages of rocks became abundantly available, in the 1950s to the 1970s, Earth history could be charted precisely. The stratigraphic column, which used to be just a record of the succession of different fossil types, could now have ages attached to it so that we could see exactly

*A violent impact or other such events can also drive off argon and "reset the clock." Thus potassium-argon ages of some meteorites or lunar rocks measure the time since that impact, not their age since formation. This is revealed by fracturing, the shock-altered structure of mineral grains, and other signs of a violent impact.

† The famous radiocarbon method of dating, based on the isotope carbon 14 as mentioned above, is another type of radioisotopic dating, but it is used on organic material from living organisms instead of minerals from rocks. It dates the time since the organism died. For example, it is used to date bones or wood tools.

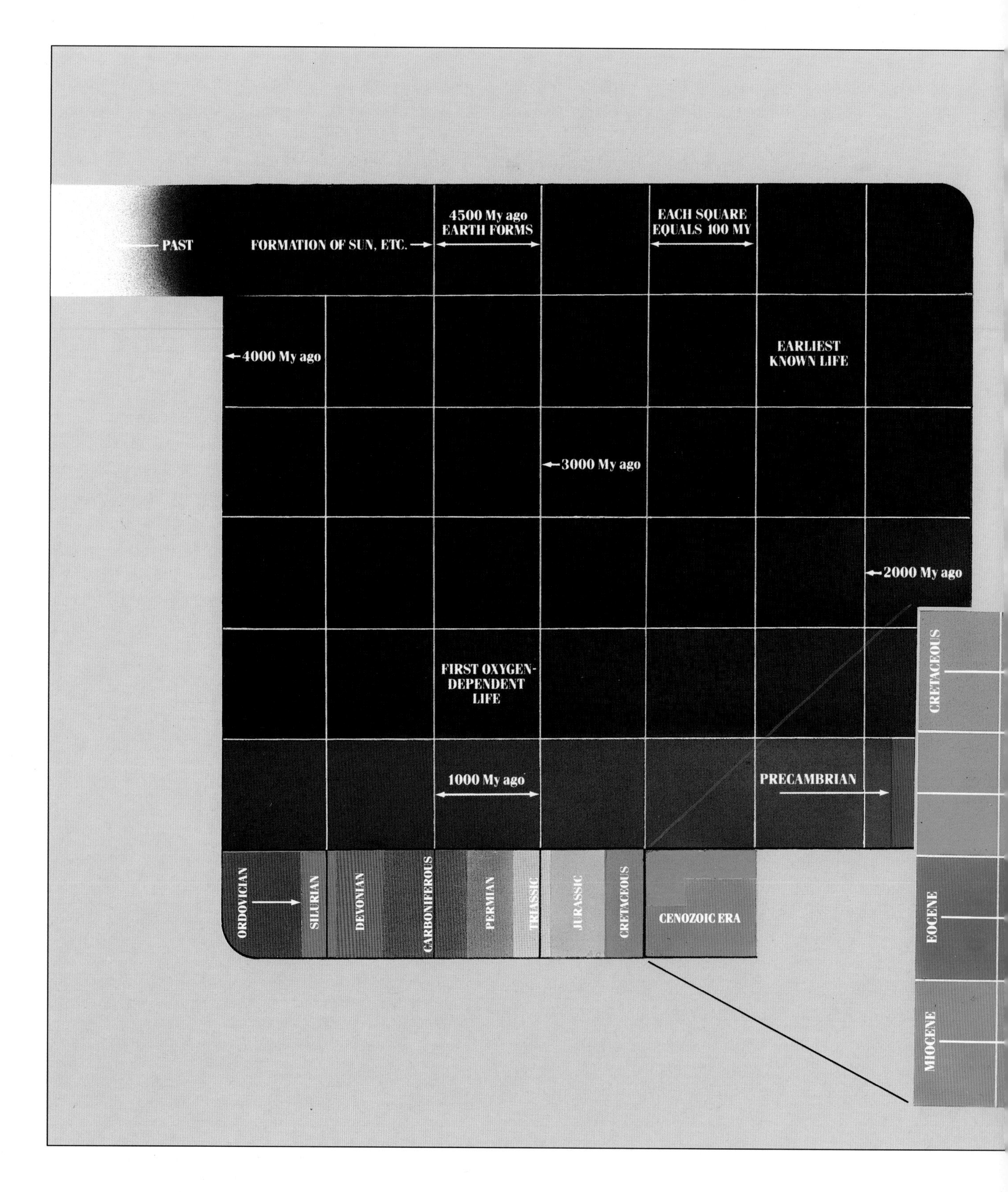
PAST
FORMATION OF SUN, ETC.
4500 My ago
EARTH FORMS
EACH SQUARE
EQUALS 100 MY
4000 My ago
EARLIEST
KNOWN LIFE
3000 My ago
2000 My ago
FIRST OXYGEN-
DEPENDENT
LIFE
1000 My ago
PRECAMBRIAN
ORDOVICIAN
SILURIAN
DEVONIAN
CARBONIFEROUS
PERMIAN
TRIASSIC
JURASSIC
CRETACEOUS
CENOZOIC ERA
CRETACEOUS
EOCENE
MIOCENE

when each type of lifeform existed and when specific mountains and continents were formed.

Construction of an accurate geologic time scale was obviously a great step forward in understanding Earth. The result is shown at left.

One of the interesting facts that emerged from establishment of the geologic time scale is that the Earth itself is much older than well-identified fossils. When Hutton, Lyell, and the others divided Earth's history into different periods according to fossils, the oldest period they could recognize—the one with the simplest fossils of deep-sea creatures—was the Cambrian. They might have thought that these "first" creatures arose only slightly after Earth formed. A Precambrian Period was inserted in the stratigraphic column to take care of whatever time elapsed before creatures evolved who could leave prominent fossils. Once radioisotopic ages were available, it turned out that the Precambrian occupied almost 90 percent of Earth's history! All the succession of periods that Werner, Hutton, Lyell, Cuvier, and the others had argued about encompassed only the last 13 percent of geologic time!

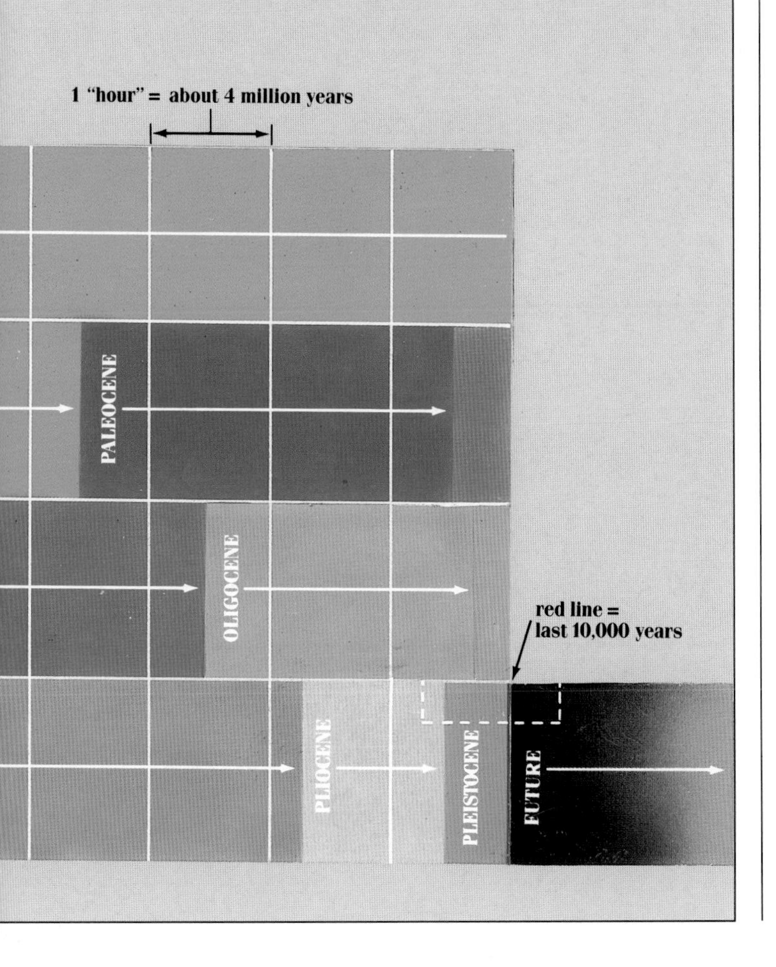

The history of the planet Earth has been reduced to this "calendar," where each "day" (represented by the individual squares in the left-hand part of the diagram) is equal to 100 million years. The last "day" (expanded into the right-hand part of the calendar) represents the most recent 100 million years; this square has been subdivided into 24 "hours," each equal to about 4 million years.

More than three-quarters of the calendar is black because we lack a good fossil record for the first three billion years of Earth's history. Only during the last "week" does life burgeon, creating the fossil record that defines the geological eras and periods depicted in color at the bottom. Mammals did not appear on Earth until the last "day." Human beings did not appear until the very end of this "day" (the end of the small blue segment in the lower right-hand corner). The last 10,000 years of human history is represented by the thin red line. The three or four thousand years of recorded history would make a line too thin to reproduce here.

The colors used here are repeated in the "timeline" that runs throughout the book.

BOOK TWO

Birth of the Earth-Moon System

4550 to 4520 My Ago

3. Before Genesis

The earth was without form, and void, and darkness was upon the face of the deep.

—*Genesis, 1:2, ca. 800* B.C.

arth is old. Earth is young. Both sentences are true because language is vague. Which sentence you choose depends on your frame of reference. In terms of a human lifetime or human history or even in terms of the existence of mammals, Earth is inconceivably old. Earth formed a hundred million lifetimes ago. Try to imagine the total length of time that humanlike creatures have roamed the Earth. Earth formed a thousand times longer ago than that. We are only a tenth of a percent as old as our planet. At 4500 My age, Earth is old.

Yet Earth is young. Earth and sun formed at about the same time, relative to the cosmic time scale, but this was long after the universe began. Earth and sun and the sun's entire family of planets are only one part of the vast Milky Way galaxy. This galaxy is a ponderous spiral of 100 million stars, and our sun is only one star among these 100 million. The Milky Way galaxy, like other galaxies, began to take shape about 14,000 My ago, shortly after the universe as we know it began in a strange, explosive event called the Big Bang. Many generations of stars formed in the Milky Way galaxy before the sun and its family of planets ever appeared. Some of those stars still exist in our sky. Others long ago dwindled into red cinders or exploded in cosmic flashbulb bursts bright enough to outshine an entire galaxy. Perhaps 8000 My of stellar comings and goings elapsed before the sun appeared as one tiny, luminous dot in the Milky Way. So Earth is only a third as old as the universe. Earth is young.

For 8000 My the stars of the Milky Way circled around the center of the Milky Way's disk, organizing themselves into a ragged pinwheel of spiral arms. The arms are the regions of stronger concentration of gas and dust—the kinds of regions that spawn new stars. At the beginning, the stars of the Milky Way were composed of almost pure hydrogen and helium, but the nuclear reactions inside stars fused these atoms into more massive ones—carbon, nitrogen, oxygen, silicon, and iron, to name a few. When massive stars exploded in supernovae, they spewed these so-called heavy elements into space, where they circulated in the form of atoms and grains of dust until they were incorporated in later generations of stars and planets. A good thing it was for us, too, since Earth is made largely of oxygen, silicon, and iron, and we ourselves utilize carbon's ability to form complicated self-replicating molecules with hundreds of thousands of atoms. Without the first few thousand My of stars cooking up heavy elements in their pressure cookers, neither Earth nor its occupants could have existed.

How did new stars, such as the sun, come into being? The clouds of interstellar dust and gas in the galaxy's spiral arms were far from static. As these clouds orbited around the galaxy and as supernova explosions blasted new gas into them, the clouds were disrupted, sometimes compressed and sometimes blown into tattered, filamentary fragments. Sometimes the clouds were compressed beyond a certain critical limit. At this limit, the atoms and grains are pushed close enough together that their mutual gravitational attraction, one for another, overcame the forces of dissipation. Even though the density of gas and dust was much less than that of the air you are breathing, the density was great enough and the cloud was big enough that gravity dominated any disruptive tendencies. The cloud for the first time had its own permanent identity, and gravitation made it begin to contract. Usually it broke into hundreds of smaller sub-clouds, each itself contracting. Each shrank and shrank until it reached the size of a star. At the same time the center of such a sub-cloud became hotter and hotter until nuclear reactions started. At that point it had become a star.

This was the process that led to the continual formation of new stars in the spiral arms, and it was the process that led to formation of the sun and Earth 4550 My ago.

We noted that star-spawning clouds generally break up into a whole cluster of stars, like the Pleiades—the famous grouping of the Seven Sisters, prominent in the autumn skies. Almost certainly the sun formed as a member of such a group. Such groupings dissipate within a few hundred My. This means that the skies of primordial Earth were ablaze with a hundred stars as bright as Venus, much closer than the stars of today's skies.

The gradual shrinkage of a contracting cloud and its breakup into stars takes a few tens of My. This may seem like a long time, but it's short in terms of cosmic time—less than a percent of the sun's history of 4550 My. The sun's formation was finished in a cosmic moment.

WHAT THE STARS TELL US

One way to find out what happened in this cosmic moment is to look at newly born stars that are forming even today. They are surrounded by clouds of debris left over from their formation—clouds of gas and dust grains. Such a debris cloud is called a cocoon nebula because it surrounds the newly forming star like the cocoon of a newly forming butterfly. The debris circles around newly born stars in a thin disk, in the same way that the debris particles orbiting around Saturn have formed rings. In some cases, these flattened disks have actually been glimpsed in photos of young stars. Spectra show that the dust grains in the disks have varied compositions, including silicate minerals, carbon compounds, and ices such as frozen water—the same materials that made planets around the sun.

The observation that young stars' cocoon nebulae contain planet-forming material was a great breakthrough because it confirmed earlier theories about how Earth formed. For years, planetary scientists had theorized that the sun's planets formed in a disk-shaped dust cloud orbiting around the primordial sun. Among the lines of evidence is the fact that the solar system as a whole has the shape of a disk. The planets, including Earth, lie in about the same plane and all move in the same direction. Theorists had named the sun's disk-shaped cocoon nebula the solar nebula. The solar nebula was the placental material for Earth's birth. Interplanetary space, at the time Earth was forming, was not as empty as it is now.

But just how did the grains and gas interact to form a world? Prior to the 1950s, this process was not understood at all. One theory was that planets formed from blobs of material torn out of the sun, perhaps by a passing star. Another theory was that they formed from contraction of large regions of the nebula. Researchers noted that if such blobs or regions were massive and dense enough, their own self-gravity (the attraction of one atom for another) would make them shrink in the same way that larger, interstellar clouds contract to form stars. They would be too small to form stars but could form planets from the size of tiny Mercury and Pluto to that of massive Jupiter. Of course, most of the solar nebula, like the sun, consisted largely of hydrogen and helium, but the atoms of these light gases are too light and fast-moving to be held by Earth's gravity. They would drift off, leaving the Earth composed of heavier elements, as we see today.

Further work, however, showed that this idea does not fit the facts. There are three major clues about what happened: the chemistry of xenon and other inert gases, physical studies of collisions among dust grains, and studies of meteorites.

Clue 1: What xenon tells us. During the 1950s, Nobel-Prize-winning chemist Harold Urey developed one of the key chemical clues about Earth's origin—a clue that Earth formed from small, rocky bodies instead of from compression of a giant cloud of solar gas. This clue is easy to understand and is still significant today. The heavy, inert gas elements, such as xenon, krypton, and argon, are typically less than a millionth as abundant in Earth as they are in the sun. However, the planets must have formed in the solar nebula, an environment with the full solar complement of inert gases. So why doesn't Earth have solar abundances of these heavy gases?

Urey noted that Earth couldn't have been formed in a one-step direct gravitational compression of a large blob of solar or nebular gas because it would then have had a full solar percentage of inert gases. These gases would have been too heavy to escape later from Earth's gravity, and we would find more traces of them today in Earth's atmosphere or soil.

The key property of inert gases is that they don't combine easily with other elements to make compounds or rock-forming minerals. Thus Urey explained this depletion of inert gases by arguing that nebular material aggregated into many small bodies, *with gravity too low to retain inert gases.* Because inert gases fail to combine into chemical compounds, the minerals in those small bodies would have lacked inert gases such as xenon. These small bodies are known as planetesimals. Ranging in size from dust grains to asteroids hundreds of kilometers across, they were the intermediate step between the solar nebula and a finished planet. They collided with each other to aggregate into an Earth that lacked inert gases.

Clue 2: What planetary orbits tell us. Around the same time that this chemical evidence was es-

4552 My ago. Earth's difficult beginnings. The process of collision among planetesimals in the early solar system was both a builder and a destroyer of worlds. While early collisions were gentle enough for colliding bodies to aggregate gravitationally and build worlds, some later collisions were violent. This view shows one of the many embryo planetesimals in Earth zone being hit and shattered by a relatively high-speed asteroid (darker body) that was deflected into Earth's zone from the outer solar system.

tablished, Soviet researchers, using different reasoning, also concluded that Earth and the other planets must have originated during a step-by-step buildup from small colliding grains called planetesimals. Soviet scientist O. Yu. Schmidt, in particular, set the stage for the understanding of planetesimals. He argued in the 1950s that the regularities of orbital and other motions in the solar system required that planets aggregate from millions and millions of smaller bodies. The regularities include such features as:

- Concentric circular orbits.
- The concentration of the orbits nearly in the plane of the sun's equator.
- The motion of all the planets in the same direction around the orbits (a direction called prograde—counterclockwise as seen from above the North Pole).
- Rotation of most of the planets in this same prograde sense around their own individual spin axes.
- The small or modest tilt of most planets' spin axes to the plane of their orbits.

Schmidt's argument was that if each planet had formed independently from massive blobs of material torn from the sun, their resulting motions could have been random: there could have been elliptical orbits, orbital motions and spins in wildly different directions, or randomly tilted axes. Schmidt and his followers reasoned that the planets must have formed from a disk of dust and gas that contained innumerable small planetesimals bumping into each other. They would have moved in circular orbits, because any particle deflected onto an elliptical orbit would cross the paths of others, bump into them, and quickly be nudged onto a less colliding path parallel with its neighbors. In the same way, if you are in a car race on a circular track, you can move along with the neighboring cars as long as all of you keep roughly parallel courses, in the same "prograde" direction, but if you try to weave back and forth across the stream or drive in the opposite "retrograde" direction, you will not last long!

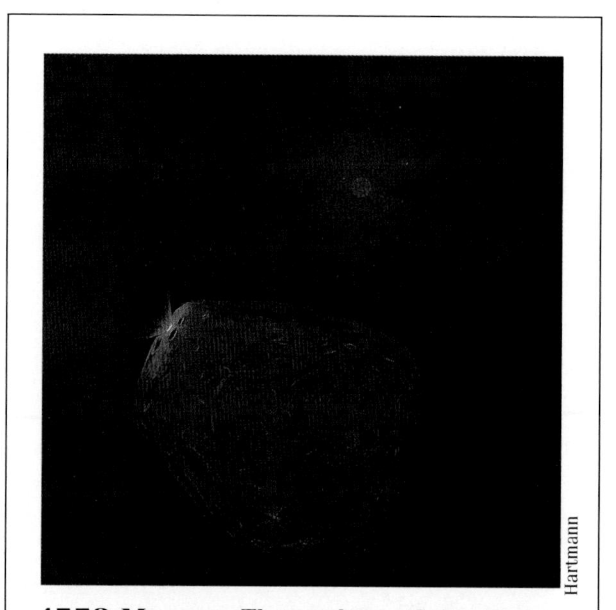

Hartmann

4552 My ago. The embryo that became Earth. During a certain point in the aggregation of planetesimals, by random chance one was larger than the rest. Its gravity caused it to attract other planetesimals from the surrounding dust clouds of the solar nebula, so that its growth became still more rapid. Here, at a diameter of 300 kilometers, the Earth-embryo is hit by one high-speed impactor. By the time it reaches a diameter of 500 kilometers, this body will have a more spherical shape due to gravitational forces. Dust clouds of the solar nebula dim the background

Soviet researcher Victor Safronov and his colleagues extended these ideas. They calculated the results when millions of particles, orbiting around the sun, began to collide. The calculations confirmed that planetesimals aggregate to produce planets in circular orbits, all in about the same plane, and with all the other properties familiar to us.

The key to Earth's growth, from planetesimal dust grain to full-fledged planet, was continued collision among the planetesimals. Because the initial grains were forming and moving together in a gaseous medium, like falling snowflakes, they did not move very fast compared to their neighbors. They collided only gently. This enabled them to aggregate, just as snowflakes clump in a blizzard.

The silicate grains of Earth's zone aggregated into bigger and bigger bodies in this way. At first they collided too fast to break each other apart. Sometimes the collision was too fast, and they would bounce apart like two rocks tossed together in the air. But some of the collisions were slow enough that the gravity of the two bodies would hold them together once they bumped into each other. In those collisions, the two ended up aggregating, making a still larger body. Thus grains grew into asteroid-sized bodies, kilometers across. Planetesimals covered a range of sizes, from BBs to mountains.

This aggregation process is called accretion. Accretion proceeded very fast, in terms of the geologic time scale. According to some calculations, the growth from microscopic grains to asteroid-sized mountains may have taken only 0.1 My or less. The process slowed then, because the material was being concentrated into fewer and fewer bodies. The solar nebula was being cleared.

By the time the planetesimals reached the size of mountains, they were colliding at higher speeds, but by then they were big enough to have even stronger gravity, which assured that most colliding bodies aggregated. Thus they continued to grow.

However, a destructive process now began. The remaining bodies were colliding at higher speeds, so fragmentation sometimes occurred instead of accretion. The cause of the higher speeds was the growth of the planets themselves. The larger a body grew, the more gravity it had. The more gravity, the greater its ability to disturb the motions of smaller bodies around it. Thus as one of the remaining small planetesimals underwent a close miss with one of the few emerging "world-class" bodies, it was sent zooming off in a different direction, rather like the Voyager space probe caroming off toward Neptune after going by Uranus. Like a car suddenly taking a sharp right turn across neighboring lanes on the freeway, it was likely to have a relatively violent collision with the next body it encountered. As the planets grew, the planetesimals' orbits got more disturbed, and the more disturbed they got, the greater the average collision velocity.

The few bodies that were winning the growth competition among the accreting planetesimals were not bothered by this problem. They were the emerging planets, the world-class bodies. They were so big that even if they were zapped by a high-speed planetesimal, their gravity trapped the ejected fragments. So each impact resulted in a net accretion of mass. They were sweeping up the remaining small planetesimals. However, the remaining smaller planetesimals were colliding with each other at greater and greater speeds, and instead of growing, they were being smashed to bits. These remaining broken fragments are the meteorites; even today, some are left, and they collide with Earth occasionally, pelting us with bits of stone and iron. They give us our third set of clues about Earth's formative conditions.

Hartmann photo

A bowl-shaped 20,000-year-old impact crater, nearly a mile across, is a reminder that major impacts can still disturb the seemingly placid Earth. Meteor Crater, Arizona.

4551 My ago. The violent process of growth. An impact on the half-finished Earth obliterates one region with a new crater, but adds mass to the forming planet.

Clue 3: What meteorites tell us. Two hundred years ago, at the time of the American Revolution, very few scientists thought that it was possible for stones to fall out of the sky. Such tales were branded as superstition until 1803, when a French village was peppered with stones, and a French Academy investigation collected samples and established that it had really happened. This marked the scientific acceptance of meteorites as cosmic samples.

Meteorites are incredibly complex rocks. They resisted easy analysis for 150 years. Finally, through sophisticated chemical analysis, microprobe techniques, and comparison with lunar rocks, they have begun to tell us their story. They reveal that they are pieces of the original planetesimals. Since the days of Earth's formation, many of them have been preserved in the asteroid belt between Mars and Jupiter, whence they are occasionally deflected toward Earth.

Meteorites reveal that by the time planetesimals had grown to sizes of a few hundred kilometers, or a few hundred miles, many but not all of them melted. This allowed the melted meteorites to form cores of nearly pure nickel-iron alloy, like Earth's core, and to erupt basaltic lavas onto their surfaces. The unmelted ones preserve more of the story of primitive conditions. Their structure confirms that the dust grains aggregated as described above.

Meteorites also confirm that planetesimals, once they grew larger, smashed into each other at higher speeds. Many meteorites are not coherent rocks but so-called breccias, or rocks made from jammed-together fragments of different, earlier rocks. Often the fragments themselves represent two different kinds of meteorites—the two bodies that smashed together. They are the fossil records of collisions in space.

The meteorites themselves give us the exact dates of all these events. Radioisotopic dating shows that virtually all meteorites formed during a relatively short time interval about 4550 My ago. More detailed studies have shown that they aggregated from smaller dust grains during an interval as brief as 20 My—an extremely short burst of planetary formation at the beginning of the solar system's very long history.

CONTINUED STUDIES OF EARTH'S ACCRETION

All three of the above-described clues—early chemical studies of inert gases, calculations of accretion physics, and the studies of meteorites—confirm that innumerable small planetesimals had to be involved in the aggregation of Earth and other planets.

The Soviet theorizing about accretion was little noticed in the West until around 1970, when Safronov's book was translated. The Soviet dynamical work, when combined with the American chemical studies, produced an explosion of new understanding of planet formation in the 1970s. Geochemists in many countries measured ever-more-precise ages of meteorites, Earth rocks, and finally, moon rocks, affirming that the whole process began about 4550 My ago. Dozens of theoretical papers detailed how the planetesimals would collide and aggregate into asteroid-sized and moon-sized bodies, within a few My. The largest "embryo" would gradually pull ahead of the rest in size. Earth might have reached half its present mass in the first million years, and then taken tens of millions of years to sweep up the rest of the debris and grow to its present size.

Some recent calculations indicate that by the time Earth reached half its present size, much of the mass of the planetesimals was concentrated not in mere asteroid-sized bodies, but in moon-sized or Mercury-sized bodies, as much as a third as big in diameter as the present Earth. As we will see, these had important consequences as they eventually crashed into Earth and other planets.

WHY IS EARTH MADE OF ROCK INSTEAD OF ICE?

The compositions of the planetesimal dust grains were controlled largely by the temperatures in the solar nebula. The nebula slowly cooled, but the inner regions near the sun were always the hottest. As the gas cooled, more and more dust grains condensed in it, just as snowflakes and hailstones condense in cooling air. In the hot inner regions, where Earth formed, only metallic and silicate grains could condense, explaining the basic rocky composition of Earth. In the direct sunlight of space, Earth was too hot for much ice on it to be stable. Farther out from the sun, carbon-rich and icy grains condensed, explaining the soot-black asteroids of the outer asteroid belt and the ice-rich compositions of the outer planets and their moons. As we will discuss in another chapter, ice-rich debris from the outer solar system may have played a role in Earth's finishing touches.

EARTH'S RAPID FORMATION

Estimating that Earth's growth rate was winding down after 40 million years, we can say that the main birth processes were happening from about 4550 to 4510 My ago—a very "rapid" and narrow interval of time, compared with the whole history of the solar system.

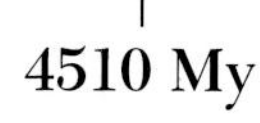

4510 My Ago

4. Earth Approaches Its Present Size

Where were you when I laid the
foundation of the earth?
Tell me, if you have understanding.
Who determined its measurements—surely you know!
. . . who laid its cornerstone,
when all the morning stars sang
together,
and all the sons of God shouted
for joy?

—*Job 38:4–7, ca. 600 B.C.?*

How long did it take to complete Earth's growth? This question is difficult to answer, but there is some evidence hidden in isotopes of terrestrial materials, lunar rocks, and meteorites. Taken together, these data suggest that Earth reached its present size only about 70 ± 20 My after the solar nebula's formation. That is, we can say that Earth started as a rocky embryo about 4550 to 4510 My ago, and reached full planethood perhaps 4510 to 4460 My ago.

What was Earth like as it reached its present size? What did it look like? What was its surface environment? For many years such questions were almost ignored by scientists because there seemed to be no way to answer them. However, new data have begun to accumulate. In the late 1980s, scientific conferences were held in locations as diverse as Oxford, Berkeley, Moscow, and Australia to discuss the new information. Sources of data range from arcane chemical studies of terrestrial and lunar rocks to orbital calculations using once-classified supercomputers.

These conferences have established a new view of Earth's early days, around 4500 My ago. Once-mysterious details are being filled in, and some surprises have been added to the once-sketchy story. Whereas scientists used to assume that Earth's primeval surface resembled a rocky desert or a global ocean, a surprising new image has emerged.

A RAIN OF ASTEROIDS

In order for Earth's mass to accumulate out of planetesimals, an incredible rain of such objects had to fall on Earth's surface as the planet grew. These objects ran the gamut from small to large, including tiny sand grains; rocky fragments similar to today's meteorites; larger rocky bodies many kilometers across, identical to today's asteroids; and gigantic bodies, even bigger than today's largest asteroids.

We can get an idea of the incredible asteroidal rain by noting that the whole planet had to accumulate in an estimated 70 My, as mentioned at the beginning of this chapter. The impact rate necessary to do this averaged a billion times today's rate! At today's rate of meteorite infall, every century or so we get an impact big enough to throw dust into the stratosphere and color sunsets around the world. But at the primordial impact rate 4500 My ago, large impacts every month or so threw enough dust into the air to block out the sunlight completely! Calculations suggest that it would take more than a month for such palls to clear. Thus one aspect of our new image of Earth's initial environment is that the meteorites struck frequently, and

Hartmann

4551 My ago. A half-sized Earth. As the growing Earth approached half its present size (about the size of Mars), it was a heavily cratered world beginning to acquire a thin atmosphere. Planetesimal impacts were very frequent, as shown by the two glowing sprays of crater ejecta, visible on the night side. Aurorae caused by the strong flow of atomic particles from the early sun play around the poles. Cloud patterns are more belt-like than Earth's present sworls because of the primitive planet's faster rotation.

their ejected dust probably kept the skies dark most of the time.

The rain of asteroids continued for hundreds of millions of years, and in Chapter 6 we will discuss some of its later consequences.

AN OCEAN OF MOLTEN LAVA

A challenge in reconstructing primordial Earth has been to reveal how the surface was affected by the shower of asteroidal debris. As we mentioned in the last chapter, the earliest planetesimals collided too slowly to generate much kinetic energy or heat. Thus, as the embryonic Earth began to grow to asteroidal sizes, it may have been too cold to melt or to be altered by heating. Its rocky and metallic minerals were mixed throughout its substance, in whatever proportions they had when they arrived. We find the same condition in certain primitive meteorites that were never heated enough to melt.

Geologists call such material *undifferentiated,* because the chemical components have not separated from one another; in particular, the metal components have not separated from the rocky components. This was the state of the embryonic Earth, perhaps 100,000 years after it began to form.

However, as Earth grew larger, its gravity grew stronger and it pulled in planetesimals at higher velocities. Impacts were now more like explosions. They released heat. The heat of a nearly continuous rain of high-speed planetesimals—i.e., asteroids and asteroid fragments—melted the surface layers as fast as they accumulated. The largest of those were especially effective. Whereas small impacts release explosive heat at the surface, where it can dissipate, the largest planetesimals made craters more than 1000 kilometers across and excavated as deep as 40 to 100 kilometers. Much of the heat released was transmitted to material at that depth and then covered by debris, so that it was "stored" in Earth's upper layers. According to calculations by some scientists, there was enough heat created by impacts to keep at least the upper tens of kilometers of Earth molten.

This means that as primordial Earth approached its present size, its original landscape was not a cratered desert or a wave-tossed ocean but rather a sea of molten lava! Scientists call it the magma ocean. The magma ocean is pictured as a deep pool of molten material, perhaps crusted over by a thin layer of cooled rock with many glowing fractures, as can be seen on ponds of lava in modern volcanoes.* Just how deep this magma ocean was is controversial. Some researchers think of it as a molten layer hundreds or even thousands of kilometers deep. Others believe it was much shallower.

The first evidence for this surprising concept came not from Earth but from the moon. Lunar rocks brought back to Earth by the Apollo astronauts and by Soviet robotic Luna probes showed that the most ancient lunar crust was made of the light minerals that float on molten magma. This type of rock layer could have accumulated only if the moon's surface was originally a vast sea of molten lava.† The term *magma ocean* was coined to describe the original lunar surface. As we will see in the next chapter, the moon probably accreted in orbit around Earth, at a faster rate than Earth. Thus it probably had a higher rate of impact heating, producing a deeper magma ocean on the moon than on Earth. But many scientists think the same process of accretional impact heating was effective enough on Earth and other planets to keep at least the outer kilometers.

*The words *magma* and *lava* both refer to molten rock, but magma usually refers to molten material underground or melted in place, while lava refers to magma that has been erupted from a volcano and has flowed across Earth's surface.

†The low-density minerals in these rocks are called feldspars. They are very common in terrestrial and lunar surface rocks. The oldest parts of the lunar crust are made of rocks composed almost entirely of accumulated feldspar crystals—a type of rock called anorthosite. The feldspars must have floated on the magma ocean and have been concentrated on the moon's surface ever since. Some scientists think that Earth's original surface may have been similar, but of course it has been altered by processes of erosion absent on the moon.

4551 My ago. The beginnings of the magma ocean. As Earth grows, its gravity gets stronger, increasing the speed of impacts. The high-energy impacts cause heating of the surface, eventually leading to widespread melting—the magma ocean.

WHAT EARTHQUAKES TELL US

Earth's interior today is very, very hot and quite differentiated into regions of different composition. There are three basic regions. At the center is the *iron core*, extending about halfway to the surface. Although it is loosely called the iron core, this region is actually a mixture of iron and smaller amounts of nickel and other materials. It is better described as a nearly pure nickel-iron alloy. The outer portion of the core is molten, but its inner portion is solid (due to the increased pressure toward the center).

Most of the rest of Earth is composed of dense rock surrounding the core and extending nearly to the surface. This part is called the *mantle*. Mantle rocks are richer in metal than surface rocks but are of course much poorer in metal than the nickel-iron core. Among the minerals believed to be common in the mantle is the dense greenish magnesium-iron silicate mineral, olivine.

Finally, the surface layer is called the *crust*. It averages some tens of kilometers thick, composed of more familiar rock types such as granite, basaltic lava, and sedimentary rocks. It is thinnest under the oceans, only about 7 kilometers (4 miles) thick. It averages 40 kilometers (25 miles) thick under the continents, which are in effect thick blocks of low-density crust floating on the mantle. It is thickest under massive mountain ranges, reaching 80 kilometers (50 miles) in depth.

The idea of a core, mantle, and crust inside Earth is not just speculation. There is direct evidence about Earth's interior structure. Seismic waves from earthquakes penetrate from the earthquake site through Earth to seismometers on different sides of Earth's globe. These allow geologists to interpret the interior structure between the earthquake and the seismometer. The results reveal a high-density central region with the density of metal.

4510 My

4540 My ago. Earth's atmosphere was at this point completely enshrouded in clouds of dust from the intense bombardment of meteorites. Beneath the clouds, crusts of cooling rock formed on parts of the magma ocean, only to be remelted by new impacts and new eruptions.

Certain kinds of seismic waves do not pass through liquids. When an earthquake is positioned partway around the world from a seismic observer, these waves would have to pass through the outer part of the core to get to the observer, but they do not make it. This shows that the outer core is molten. The inner part of the core, though hot, is solid, because the central pressure is high enough to force the inner core into solid form. So Earth's center is occupied by a giant ball of nickel-iron alloy, molten in its outer layers and solid at the center.

MAKING EARTH'S IRON CORE

The draining of iron to Earth's core and the resultant creation of the iron-depleted mantle together make up one of the most significant processes in Earth's history. After all, the mantle com-

poses about two-thirds of the planet's mass, with the core composing most of the other third. How and when did the mantle and core take shape?

The very presence of an iron core indicates that Earth has gone through enough heating to allow the iron to separate from the rock. The mantle did not have to get completely melted for this to happen, though some researchers think it was once completely melted. Even if the temperature only got close to the full melting temperature, iron blobs could have formed (containing nickel and other impurities attracted to iron), and these would have been dense enough and heavy enough to force their way downward through the partly melted mixture of magma and silicate crystals. The hypothesis of a magma ocean clarifies how this may have happened. As Earth grew, impacts not only continued adding material, layer by layer, but also kept the surface magma ocean molten, probably for millions of years. Maintenance of the material at or near the melting temperature allowed the heavy, molten blobs of nickel-iron to drain through the hot, plastic interior toward Earth's center as fast as the planet formed. This idea is a topic of intense current research (for example, see the review by Newsom and Sims, 1991, listed in the references).

The process resembled that in a smelter vat of molten or partly molten ore. When iron ore is melted in the vat, molten iron sinks to the bottom and a silicate-rich rocky slag floats on the top. In a sense, primordial Earth was like a giant smelter vat, with the iron at the bottom in the core and the silicate-rich, lower-density rocks at the top in the crust.

At the surface, as the impact rate declined, the magma ocean cooled and lightweight minerals collected there. Thus, as time went on, Earth accumulated a thicker crust as well as a better-defined iron core.

This theory of early core formation *as Earth formed* is a radical revision of the theory of only a decade ago, according to which Earth formed in a fairly cold state and heated only *after* it reached full size. Scientists once assumed that radioactivity alone was responsible for the heat. Most radioactive elements in natural rocks release heat only slowly. In the older theory, therefore, Earth started relatively cool and heat built up only slowly, over hundreds of millions of years. The center got hot fastest, because heat could not escape. The upper mantle did not heat as fast, because heat was conducted out to the surface and radiated into space.

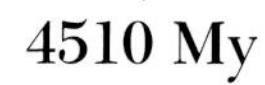

4540 My ago. When Earth reached its present size, the solar system was still embedded in the star cluster from which the sun itself formed. Many bright stars filled the sky. A thick atmosphere shrouds the planet.

In this older theory, the core formed hundreds of My after Earth formed, when inner Earth finally reached the melting point, and iron drained rapidly (on a geologic time scale) toward the center in an "iron catastrophe" that was thought to have changed Earth's mantle structure and spin rate.

In the new theory, the iron core was forming continuously during the first 10 or 20 My or so, as Earth grew. As we will see in the next chapter, this rapid formation of the core is needed to fit modern evidence about the origin of the moon.

If the new theory of impact heating and rapid core formation is correct, then what about the role of heat produced by radioactivity? Radioactive elements are inside Earth in any case, and did add their heat. The outer layers of Earth insulate the center and keep the heat in. Even during the entire history of Earth, there has not been enough time for that heat to escape from the central regions to the surface. The buildup of radioactive heat released in the central regions over the eons has kept part of the iron core molten even today.

An important facet of core formation is that it was not necessary to heat the original mantle to a full molten condition in order for the iron to drain out. As the rock approached molten conditions, it acquired a plastic quality, with some minerals melted and others still solid, so that the iron core could form without the mantle being fully molten. Radioactivity may never have made the mantle much hotter than that.

According to a few researchers, as the iron drained to form the core, it may have trapped the primordial rocky embryo in its interior. These researchers claim to see evidence of volatile materials from the primitive embryo still leaking out from the core into the mantle, though this view is controversial.

WHAT MANTLE ROCKS TELL US

Unfortunately it is hard to test these theories because the mantle is covered by the crust, some 5 to 50 kilometers deep, so that we have little direct access to mantle rocks. Luckily, a few rocks from the upper mantle are occasionally ejected by volcanoes that draw their magma from deep roots, more than 100 kilometers down. The best way to study the history of Earth's mantle has been to make very careful studies of the chemistry of such upper mantle rocks.

They can be compared, for example, with primitive unmelted meteorites, which give a fair idea of Earth's primitive composition before it melted and differentiated. Upper mantle rocks are nowhere near as iron-rich as unmelted meteoritic rock. Meteorites indicate that Earth's primitive building blocks had three to four times as much total iron as Earth's mantle has today, confirming that iron (and other elements that tend chemically to associate with iron) drained out of the mantle in order to form the core. The mantle is depleted in these iron-affinity elements. To a first approximation, then, mantle composition makes sense to researchers.

However, scientists are never satisfied with first approximations. As their instruments get better, they measure the abundances of different elements with ever-greater accuracy, and to no one's surprise they find new puzzles. Two major puzzles are, first, that the upper mantle has too many oxygen-rich compounds to have been in equilibrium with the iron draining to the core, and second, it has somewhat more iron-affinity elements than expected, since these should have drained efficiently to the core.

These puzzles are now believed to offer clues about the history of the mantle during its formation. For example, in the early 1980s, German geochemist Heinrich Wänke proposed that the composition of material falling on Earth changed slowly *as Earth was forming*. In this view, the last 20 percent of material had more oxygen-rich compounds, which never fully mixed with the rest of the mantle or core and thus left the upper mantle richer in oxygen compounds and iron-affinity minerals.

The puzzle of the mantle's iron abundance was complicated in the late 1980s, when astronomers discovered that the sun has more iron atoms than had been thought. This confuses the question of how much iron primordial Earth should have acquired as it formed. It illustrates how astronomical discoveries have a habit of reflecting back onto our understanding of Earth. Geochemists are still arguing over the issue of iron abundance in the mantle.

WHAT ASTEROIDS TELL US

Still more evidence about iron cores comes from the ever-informative meteorites. As remarked in Chapter 3, the parent bodies of most meteorites were the original planetesimals, some surviving today as asteroids up to 1000 kilometers across. The meteorite data show that while some of these planetesimals—or at least parts of some of them—never melted, others did melt and formed iron cores, just as Earth did. Collisions between the planetesimals broke some of them open, exposing the cores. And some of the meteorites that fall out of the sky today are pieces of pure nickel-iron alloy, presumably fragments broken from the cores of these asteroids. Other meteorites are olivine, presumably from the mantles of these broken worlds. An interesting additional class of meteorites, called the stony-iron meteorites, are intricate mixtures of pure metal and olivine rock—presumably telltale samples from the actual core-mantle boundaries of broken asteroids. Finally, a few meteorites are lava-like rocks presumably from the relatively thin asteroid crusts. In short, meteorites indicate that at least some asteroids, though much smaller than Earth, went through similar heating and developed similar core-mantle-crust structures.

As a clinching line of evidence, telescope data on asteroids show the same pattern. We can't see the surface details of asteroids, but their spectra reveal their approximate composition. Some asteroids still have unmelted surfaces. But among the rest, we see the same telltale rock types mentioned above. One major class of asteroids appears to be metal. Furthermore, as my colleagues at the University of Hawaii, astronomer David J. Tholen and NASA astronomer Dale Cruikshank, and I first established in 1984, an olivine or olivine-metal class of asteroids offers the first direct observational evidence of mantle fragments among asteroids. Finally, a few asteroids show a basaltic lava composition that apparently represents crustal materials.

The dating of meteorites shows that most of the melting of these bodies occurred within 100 My after they began to form, not hundreds of My later. Dates of melting or strong heating, measured by several techniques in a group of seventeen meteorites, ranged from 4520 to 4420 My ago, with uncertainties of about 30 My and an average age of 4480 My. (For more detail, see the reference list in the back of the book for a paper by Turner, 1988.)

The cause of the heating among asteroids is controversial, but whatever the cause, the asteroids and meteorites give abundant evidence that many early planetesimals in the inner solar system melted and formed iron cores very soon after they were formed.

ADDING ALIEN WATER TO EARTH'S RECIPE?

As mentioned above, many geochemists believe that the last planetesimals hitting Earth were rich in volatile chemicals—that is to say, in the kinds of chemicals that would be released as gases and liquids during slight heating. The most important example of a volatile substance is ice, which of course releases water upon heating.

In the last chapter, we discussed the fact that planetesimals in the inner solar system near Earth's orbit had few volatiles; for example, that region was too hot to permit ice to form, although some water molecules may have been trapped in the minerals that were forming near Earth's orbit. At any rate, embryo-Earth, as it reached half or three-fourths of its present size, probably had less water initially than, for example, the next outward planet, Mars.

However, planetesimals in the outer solar system had lots of ice. As Jupiter and the other giant planets grew in that region, their gravity became so strong that they started to disturb the motions of their neighboring icy planetesimals, the comets. Some comets hit Jupiter, but many others came close to Jupiter and were deflected into the inner solar system from their original paths out of the outer solar system. Thus, from the point of view of a planet in the inner solar system, a shower of outer-solar-system comets began to be added to the mix of bombarding bodies, just as the planets finished growing. The late stages of Earth's growth may have witnessed showers of icy comets crashing onto our planet. Many scientists think that although some of Earth's water may have been trapped in its original minerals (a form known as water of hydration), some of it was added by comets during the final growth stages.

Geochemists can tell from studies of isotopes that most water in the oceans today has been recycled through Earth's materials, and its H_2O molecules contain oxygen atoms that were originally part of Earth's inner-solar-system building materials, not directly from outer-solar-system comets. However, as the magma oceans began to cool, some of the first lakes and seas may have been rich in comet-descended water; it is interesting to think that much of the water in the fledgling oceans may have been from the remote solar system, plastered on Earth's outer layers like the final touches a builder puts on a house.

This may explain why the upper mantle might have more oxygen than the lower mantle: the late-arriving planetesimals were more oxidized and volatile-rich. Here we have a nice example of how seemingly esoteric research in two different areas—mantle structure and comets—seems to suggest the same conclusion.

Research on primordial Earth continues. Perhaps the three most important questions about Earth as it reached its present size are (1) the existence and depth of the magma ocean and its role in forming the crust; (2) the effects of intense impacts; and (3) the timing and mechanism of core and mantle formation. All ideas about the first ten My of Earth history are still in their early stages. During the next decade, we can expect more progress in understanding Earth's infancy.

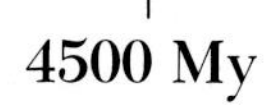

ca. 4500 My Ago

5. Where Did the Moon Come From?

> The best models for lunar origin are the testable ones, but the testable models for lunar origin are wrong.
>
> —*Lunar researcher S. Ross Taylor, paraphrased by geophysicist Sean Solomon, at Kona, Hawaii, Conference on Lunar Origin, 1984*

If you studied Earth through a telescope from space, you would see that it is different from any other terrestrial planet. It is not a single planet but a double planet, with the smaller unit being a moon one-fourth Earth's size. Mercury and Venus have no such moons; in fact, they have no moons at all. Mars has two tiny moonlets the size of Manhattan Island, but they are thought to be asteroids from the nearby asteroid belt, captured into orbit around Mars.

The giant planets of the outer solar system are also different. They have whole systems of satellites that range in size from smaller than our moon to bigger, but all these satellites are much smaller than the giant planets themselves; these systems seem to have formed by the same process that produced the family of planets around the sun.

So the question is left: how did Earth gain a solitary large satellite a quarter of its own size?*

THREE EARLY THEORIES

Three theories of lunar origin were proposed by pre-Apollo scientists. In the days before sensitivity to feminist issues, one (male) NASA scientist named these the sister, daughter, and pickup theories. According to the first, the moon grew at the same time as Earth, alongside it, as a true sister planet, starting as a tiny, Earth-orbiting planetesimal and gradually accumulating some of the debris falling toward Earth. The second theory was that the Earth got to spinning so fast that it became unstable and spun a blob of material off its equator, which became the moon. The third theory was that the moon formed somewhere else in the solar system and then was captured intact into an orbit around Earth.

Solving the mystery of lunar origin was billed as one of the main scientific justifications for sending Apollo astronauts to the moon. These explorers were trained to look for "genesis rocks"—rocks as old as the moon itself. When they brought back the genesis rocks, so the geochemists argued, careful testing of the samples would determine which of the three theories was right. We would soon learn the secrets of how the moon formed.

The first two lunar landings occurred in 1969, and the Apollo program ended in 1972 after six landings. In the intervening years, the Soviet Union flew three automated sample return missions. By 1972 we had samples from nine sites on the moon.

Early optimism about solving the problem of the moon's origin began to fade as soon as the first lunar rocks were studied. There were lots of 3000- and 4000-My-old rocks lying around, but *no* genesis rocks from 4500 My ago. The intense bombardment by planetesimals and meteorites during the first few hundred My of the solar system had pulverized virtually all rocks older than 4000 My. Still, the moon gave a better window into the primeval past than did Earth, where active erosional processes destroyed most rocks older than a "mere" 1000 My.

By the end of the Apollo program, the situation was still more frustrating. Careful studies of the nine suites of lunar rocks had ruled out all three theories! A major problem with the first theory is that while Earth has a lot of iron (mostly concentrated in its core), the moon has very little. How could the moon have grown next to Earth and magically missed out on its share of iron? Similarly, the moon has virtually no water or other volatile materials. If it grew alongside Earth, why didn't it get its share of those?

The second theory suffered from problems with dynamics. Earth really didn't have enough energy to spin up to the point of throwing off the moon, and the moon does not lie in an orbit over the equator. Furthermore, angular momentum properties of the total system today don't match the much greater angular momentum needed to spin off the moon, and the loss of the extra angular momentum was not well explained.

The third theory was dashed when it was learned that materials from different planets and asteroids have different mixtures of oxygen isotopes. The moon's oxygen isotope mixture is exactly the same as Earth's, apparently proving that the moon formed from the same basic material as Earth, despite the iron and water deficiencies.

Here was a major embarrassment! All three leading theories seemed to be shot down in flames, and there was no new alternative. One leading scientist quipped that the moon must not really exist, since our theories did not explain it.

The rocks did give some clues. Dates of the rocks revealed that the moon was not a latecomer.

*The same question might be asked about Pluto, which also has a satellite that is large compared with itself. The same theory that will be discussed below for forming the moon may apply to Pluto. But Earth is the only full-fledged double planet; Pluto itself is smaller than our own moon! It is so small that it is being downgraded by some scientists from its status as a planet to the status of a sort of glorified double asteroid.

For example, dating of one of the oldest samples of lunar crust, by geochemists Richard Carlson and Gunter Lugmair, showed that the lunar crust is 4440 ± 20 My old. This means that the moon must have formed and its magma ocean must have cooled enough to create a solid crust by that time—within 100 My of Earth's formation.

CLUES FROM THE FINAL GROWTH STAGES—EARTH ACQUIRES ITS SPIN

What was happening on Earth at that time, during its final stages of growth? A decade ago, scientists thought that the astronomical part of the story was over after the first 20 to 50 million years, and that the rest of Earth's evolution was simply a story of geology—processes such as magma cooling, crust formation, and volcanic eruptions, all operating from the inside. But now we come to one of the most revolutionary results of planetary exploration—a result that during the 1980s changed our conception of our planet: Earth was being bombarded by planetesimals from outside its atmosphere, and some of these were big enough to have catastrophic effects.

Chapter 3 discussed the work of O. Yu. Schmidt, describing how Earth accreted from many small planetesimals. As far back as the 1960s, Schmidt's student, Victor Safronov, had pointed out an especially interesting aspect of planet aggregation. As the countless small bodies accreted into one "primary" planet, a few additional big "secondary" bodies must have grown in the same vicinity. You cannot expect all planetesimals to have crowded into one planet-embryo, without competing planet-embryos starting nearby. For example, in the time required for some of the planetesimals in a given region to collide and aggregate into a body 1000 kilometers across, others in the same region might have aggregated into a body 500 kilometers across. We can actually see this situation "frozen" in the asteroid belt today. The largest asteroid, Ceres, is 1020 kilometers across. The second largest, Vesta, is 550 kilometers across. The third largest, Pallas, is 540 kilometers across, and there are many more in the diameter range of a few hundred kilometers.

In general, the second-place finishers in the accretion race ended by crashing into the winners. In other words, the planets endured not only a multitude of small impacts but a few giant impacts as well. Safronov was interested in the competition between the effects of the few biggest impacts and those of the thousands of small impacts. He noted that most of the impacts on the planet were from small planetesimals, like thousands of BBs hitting a basketball. Their net statistical effect was to produce certain "regular" patterns, such as zero axial tilt and prograde rotation.

This is how Earth got its spin in the first place. All the planetesimals and resulting planets move in the counterclockwise direction around the sun as seen from the north star—a direction called *prograde*. The net statistical effect of small bodies moving around the sun in the same direction was to accrete into planets and planetesimals all spinning in the same, prograde direction with average rotation periods of around five to twenty-four hours. This would be hard to explain by coincidence alone, but easily explained if planets grew from thousands of impactors, all orbiting around the sun in the same direction.

Safronov noted a more subtle and interesting effect. When the few moderate-sized secondaries hit the primary planets, they would be big enough to produce random "fluke" irregularities, such as tilting the axis of the planet. Safronov suggested, for instance, that the weird axial tilt and retrograde rotation of Uranus was caused when Uranus was hit late in its growth by a secondary planetesimal several percent as massive as Uranus itself.

Thus we would expect most planets to show a general pattern of regularity, like similar rotations, but we would expect that a few planets might not fit the pattern. In the words of the most famous geochemist of the last generation, Harold C. Urey, na-

Hartmann

4500 My ago. Half an hour after the collision that produced the moon. A giant planetesimal crashed into Earth perhaps 20 million years after it formed, blowing off an incandescent cloud as bright as the sun's surface. Some of the debris from the mantles of both bodies was ejected into a swarm in orbit around Earth, where it aggregated into the moon, whose composition resembles that of Earth's present mantle. The painting is based on computer simulations of the collision by University of Arizona researcher Jay Melosh and his co-workers.

ture is untidy. The countless small bodies led to the general tidiness of the solar system, but the collisions among the few large bodies led to statistical flukes that created the untidiness of the system. Or, in the words of Harvard geophysicist Ken Goettels, there was a competition in the planet-forming process between being "nice" and being "nasty."

CONCOCTING A NEW THEORY FOR THE MOON'S ORIGIN

There were rapid declines in NASA funding after Apollo, so in the mid-1970s most planetary scientists turned away from the problem of lunar origin altogether, and the whole subject languished. Quite content to go unexplained, the moon kept orbiting Earth and adding romance to evening skies.

I would like to give a more personal account of the next part of the story, because I was involved in suggesting a new theory that came to be the most widely accepted current idea of the moon's origin. As a graduate student in the 1960s, I had been reading some of the Russian work on aggregation of planetesimals to grow planets. (Graduate student years offer that rare period in a scientific career when one still has time to read other people's scientific papers instead of serving on committees and going to meetings.) Thus I had become interested in the effects of cratering and of "untidy" large collisions even before an English translation of Safronov's work was published in 1972, and before it received acclaim in the West in the mid-1970s.

In 1974, Donald R. Davis and I, at the Planetary Science Institute in Tucson, made calculations on the rate at which secondary bodies could accumulate near Earth's orbit during the 30 My or so that Earth itself was accumulating. Regarding Earth as the primary body forming in our zone of the solar system, we asked what was the largest secondary body to hit it, the second largest, third largest, and so on? Already we knew from sizes of the largest lunar impact basins that bodies as large as 150 kilometers in diameter had hit the moon after it formed. We concluded that the largest body fated to collide with Earth could have been much bigger. By the time Earth was full grown, we found, secondary bodies could plausibly have grown a fourth or even half as big as Earth itself.

There was no guarantee that the biggest secondary body would hit Earth. It might have passed by at a close distance and had its orbit changed by Earth's gravity, causing it to go careening out toward the region of Jupiter, where it might have hit one of the giant planets or even been thrown by Jupiter's gravity clear out of the solar system. Still, there was a chance that one of the larger secondary bodies would hit Earth as it finished growing. While we agreed with the general idea that Earth formed *mostly* from tiny meteoritic or asteroid-sized planetesimals, we were saying that the biggest impactor could have been as big as the moon, or even bigger! In other words, Earth might have suffered a truly catastrophic impact during its early history.

We pointed out that the lunar rocks supported this. While the composition of the moon is unlike the average composition of Earth as a whole (for example, the moon has much less iron), it is remarkably similar to the composition of Earth's rocky mantle. Our idea was that the moon formed from material blown out of Earth's mantle by impact of one of the giant secondary planetesimals. Lunar rock chemistry also indicates that the lunar material was once strongly heated, which could explain why it is devoid of water and other volatile materials. Furthermore, the exact match between lunar oxygen isotopes and Earth's oxygen isotopes, just being recognized at that time, strongly indicated that the moon itself was formed from Earth's material, or at least from material formed at Earth's distance from the sun.

Our new theory of lunar origin was radical when we published it in 1975. It required not only that a giant impactor—one of the largest bodies to grow in Earth's part of the solar system—collided with Earth, but that Earth's iron core had already been formed at that early time. Let's look at this point in a bit more detail. While the moon has much less iron than Earth's average bulk composition, it has a little more iron than Earth's mantle. This would fit with the idea that the big impact happened after most but not all of Earth's core formed. Earth's core got its finishing touches after the moon formed, which makes sense, because iron draining would have been very efficient after the mantle was heated by the impact. Furthermore (as was pointed out by later researchers), the impactor's iron core must also have already formed. With the iron safely out of the way, hidden in the cores of both bodies, the debris splashed out by the giant impact would have been strongly heated mantle material—just the sort of material needed to explain the moon.

We suggested that mantle debris blasted out by the impact formed a transient cloud around Earth, and then aggregated into the moon. Later calculations indicated that such a cloud would collapse within a few months into a flat ring around Earth, similar to Saturn's ring. After that, the outer part of the ring would re-aggregate into a satellite. Thus Earth could have acquired a satellite with just the composition actually observed in lunar rocks.

The accretion time for the moon in orbit around Earth was much shorter than the accretion time for Earth in orbit around the sun. One reason was that the world-building debris was crowded into a much smaller volume. Calculations suggest that in contrast to Earth's growth period of 70 My, the moon may have grown in only thousands of years or less. This is the reason that, as mentioned in the last chapter, the moon may have experienced a higher rate of impact heating and thus may have acquired a deeper magma ocean than Earth.

One of the best features of the giant-impact theory of the moon's origin is that it explains why only one planet out of eight (or two out of nine if you count tiny Pluto and its moon) has a satellite of appreciable size relative to the planet itself. The giant impact was a statistical fluke. Whether a growing planet gets hit by a large planetesimal and forms a moon is a matter of chance. It depends on how big the largest nearby secondary planetesimal grows before it hits the planet; whether it hits early or late in the planet's growth; whether it hits tangentially or dead-on or perhaps has a near miss and gets deflected gravitationally, going on to hit some other planet instead. Other outcomes are likely. For example, a glancing hit at the pole might change the axial tilt of the planet without producing a moon. Or a glancing impact on the equator might speed up or slow down the spin rate without producing a moon. In short, the giant-impact theory helps explain various types of untidiness among the planets.

THE NEW THEORY CATCHES ON

I first presented our idea at a 1974 conference on satellites at Cornell University. During the question period after my talk, A. G. W. Cameron, a formidable expert in the field of planet formation, stuck up his hand. As a relatively young Ph.D., I feared that he would toss off some devastating critique that would blow our nice theory to smithereens. Instead, he remarked that he and dynamicist William Ward had been working independently on essentially the same idea. It was a beautiful example of scientific dovetailing. They had come at it from an entirely different direction. They had asked themselves how big a body would be needed to deliver the total angular momentum of the Earth-moon system if today's observed total angular momentum had been delivered essentially during one big impact. They concluded that the impacting body was as big as the planet Mars! Our paper on the idea was published in 1975, and they followed with an abstract in 1976.

Then a curious thing happened to the idea. In those days I had the naive view that if you have a good scientific idea, you publish it and the appropriate people will read it and start working on it. Not being a geochemist or dynamicist who could carry the idea much further, I went on to work on other problems, waiting to be crowned with laurels if our idea turned out to be any good. Instead, the idea sank out of sight. In the real world, if you have a good scientific idea, you need not only to publish it but also to go to meetings, talk about it with colleagues, and give seminars at different institutions.

When the theory began to get attention, it was criticized as being ad hoc. Critics felt we had conjured up our impactor out of thin air, so to speak, just in order to explain the moon. The idea was said to be too catastrophic. Other, more evolutionary theories should be pursued first, critics said, before such an ad hoc event could be considered. On the contrary, I claimed that the fluke-like quality of the impact was one of the strongest points of our theory. In an otherwise orderly solar system, it alone explains why we have an occasional big axial tilt (Uranus), one planet that is hardly rotating at all (Venus), a satellite system that seems to have been disturbed (Neptune), and one or two unusual moon systems (Earth and perhaps Pluto).

The early reaction against the catastrophic aspect of the theory was interesting because of its roots in the old argument between catastrophist and uniformitarian theories. All of science has been affected by the triumph of slow and steady uniformitarian theories over catastrophist theories, as described in Chapter 1. The idea of the moon being made in one big spectacular splash, instead of in a series of modest events, had a bad ring to it. Furthermore, astronomers had been burned a decade earlier by the controversies over the popular books of Immanuel Velikovsky (1895–1979), including *Worlds in Collision*. In order to explain various legends, including the biblical stories of the flood and the parting of the Red Sea, Velikovsky postulated that major collisions had been happening in the solar system even during the last few thousand years. Astronomers could prove that the solar system was not so chaotic in the last few thousand years, but despite a hundred debunkings by astronomers, Velikovsky's books had become bestsellers. A new theory attributing planetary features to giant catastrophes therefore faced an instinctive negative reaction from the scientific community.

In spite of the skepticism, a few researchers did look at the idea in more detail. Ward and Cameron published an abstract in 1978 on the evolution of the disk of debris, and Cal Tech researcher David Stevenson (later to win a prize from the American Astronomical Society as an outstanding young researcher in planetary science) published abstracts in 1983 and 1984 about the aggregation of the debris. But there was no flood of papers, and we had little sense of whether or not our idea would survive intense scientific scrutiny if lunar origin ever again became a hot topic.

A new twist in the story came in 1984. In that year, I was asked to help convene a new conference on the origin of the moon. It had been twelve years since the end of Apollo, and there was still no widely accepted theory of lunar origin in sight. The organizers of the conference felt it was time to encourage scientists to reexamine lunar data and to see if any consensus could be reached on where the moon came from. Much to the gratification of my ego, the papers submitted for the conference program revealed that a majority of the leading researchers in this field were concentrating on the giant-impact theory.

At the conference itself, to forestall the usual criticism that giant impact was an ad hoc idea and to reduce any bias against catastrophic impacts, I gave a talk called "Stochastic Is Not Equal to Ad Hoc." The point of this talk was that stochastic—that is, random—large impacts must have been a feature of the early solar system. We cannot predict which planets had the most consequential impacts, but that is no reason to dismiss the whole phenomenon as ad hoc. Other researchers gave detailed chemical and physical models indicating that a sufficiently catastrophic impact could have blown off debris with the right composition to explain the moon. After the conference, several journals reported that the giant-impact theory had suddenly emerged as the new theory of choice to explain the origin of the moon. In fact, some scientists in a 1986 volume summarizing the conference felt cautioned against a bandwagon effect that might make the theory *too* popular. By the late 1980s, the idea that giant impacts controlled certain features of Earth and other planets was revolutionizing studies of our planet's early history.

WHAT HAPPENED DURING THE GIANT IMPACT

The worth of a new theory is not established by its winning a popularity contest. A theory is good only if it inspires some useful new efforts and if it has a chance of being tested or disproved. The main work inspired by the theory since 1984 has been a set of extraordinary calculations, using giant computers and classified military explosion-modeling software to examine what would happen if two planet-sized bodies really did collide 4500 My ago. Two teams used restricted-access computers at Los Alamos and at Sandia Labs, in New Mexico, originally designed for military applications. One team involved A. G. W. Cameron, who had independently developed the original idea of the giant impact. The other team involved geophysicist and cratering expert Jay Melosh of the University of Arizona. The results suggest that the kind of collision needed to produce the moon was a partially glancing blow by a Mars-sized body. More remarkably, the results include "movies" of the collision, derived from com-

4500 My ago. Within a thousand years after the giant impact, the moon was forming. This view is from the red-hot surface of the forming moon. The sky is dominated by a red-hot Earth, recently remolded and melted by the giant moon-forming impact. The ring of debris is aggregating into countless moonlets, which in turn are eventually accreted by the moon.

puter images at different stages of the impact. It was the biggest catastrophe in Earth's history.

What an event! Perhaps 50 My after Earth began to form—that is to say, very close to 4500 My ago—a giant planetesimal approached Earth, appearing first like a bright star in the distance and then growing bigger. During the last few days before the collision it grew big enough to appear as a disk, and during the last day it looked bigger than the moon looks to us.

The first actual contact during the impact was between the surface layers of the two bodies. The crust of each was insignificantly thin, and so in effect the mantle of the impactor plowed directly into the mantle of Earth. The impact sheared off part of Earth's upper mantle and destroyed the impactor's mantle. The debris ejected into space was primarily a mixture of the two mantles, perhaps with more of the impactor's mantle. Most of the debris fell back onto Earth, but some of it sprayed into orbit around Earth.

The blasting open of Earth's mantle took an hour or more to unfold. According to orbital velocities, the impactor would have slammed into proto-Earth at perhaps 10 or 15 kilometers per second. This seems fast, but at these speeds the Mars-sized impactor would have taken ten minutes or so to move its own diameter. If we could have witnessed it from a position in nearby space, the impact event would seem to have taken place as if in slow motion. The zone of contact between the bodies looked nearly as bright as the sun's surface. An incandescent spray of debris from the mantles of both Earth and impactor unfolded slowly, spraying out hot, iron-poor silicate gases at temperatures of more than 2000° Kelvin! Much of the mantle on the impact side of each body was blasted away, exposing the iron core. According to the computer "movie" results, the distended iron core of the impactor lost so much speed that it plowed into the partly exposed Earth core, eventually being assimilated by it. Earth, flowing like molasses, soon resumed its spherical shape, with iron core and rocky mantle gradually reestablishing themselves. The iron atoms of the impactor are now buried near Earth's center, and its melted rocky material has been mixed throughout the mantle.

More interesting to us is the fate of the debris blasted into orbit. According to the computer models, some knots of material may have formed in orbit in the first hours, but these were probably not as large as the moon. Fine, hot dust continued to condense out of the remaining cooling gas. The debris formed a dispersed cloud of hot dust and gas around Earth in the first days after the impact. During the first week, this debris gradually settled into a very thin ring, similar to Saturn's. According to studies by David Stevenson, some of the material in the ring may have been briefly in molten liquid form, like a ring of lava. Once the ring formed, it began to aggregate into individual moonlets. The largest of these swept up the rest very rapidly, in terms of astronomical time scales, perhaps within a few thousand years. Earth at last had a moon.

The moon must have formed within a few diameters of Earth, where the debris was concentrated. At that time Earth looked truly like a double planet, with its moon only a few Earth-radii away, occasionally casting a giant shadow on Earth, and with the last vestiges of a tenuous debris ring still circling Earth.

WHAT THE MOON'S MOVEMENTS TELL US

Gravitational forces between Earth and the moon work in peculiar ways. They distort the shape of both Earth and moon, raising a so-called tidal

4490 My ago. Soon after its formation, the moon was located only a few Earth-radii away from Earth and cast a startling shadow on the planet. A residual ring of debris from the giant impact still circles Earth. In the background can be seen dramatic nebulosity, because the early solar system was still embedded in the star-forming region that spawned the sun. Also in the background are several comets, since icy planetesimals were being scattered out of the region of the giant planets in the first few hundred million years of solar system history.

bulge in each. Instead of a spherical Earth attracting the moon, we can think of the situation as a spherical Earth experiencing a stretching force toward the moon, which produces a tidal bulge on each side of Earth. Interestingly, the gravitational forces of the tidal bulges themselves then affect the subsequent motions of the moon. These forces, sometimes called tidal forces, cause the moon to move away from Earth. This effect can be calculated; indeed, it has been measured on the present-day moon, which is slowly receding from Earth at a rate of about an inch per year.

This means that if you run an imaginary movie of the moon's motions backward in time, you see that long ago it must have been close to Earth. This is another success of the giant-impact theory, which predicts that the moon formed from the inner debris within a few radii of Earth. Once the moon started to move outward, it swept up any remaining debris in the outer part of the ring. Any leftover de-

Hartmann

4460 My ago. The moon has moved farther out from Earth and is beginning to sweep up the rest of the ring of debris left over from its formation.

bris, closer to Earth, gradually dissipated or spiraled inward to crash on the primordial Earth, so that no trace of the ring is left today. Within a few million years, the moon was already spiraling outward from Earth, toward its present distance.

There is another important effect of the gravitational forces between Earth, moon, and tidal bulges. As the moon spirals outward, Earth's rotation slows down. The day gets longer. This means that when the moon was close, Earth rotated much faster than today's twenty-four-hour period. The change of rotation rate as a consequence of the tidal forces is well understood and can be calculated. Indeed, Earth's rotation can be measured to be still slowing down.

The striking result is that, according to calculations involving the angular momentum, Earth's rotation period just after the moon formed was only about five or six hours. A day on primordial Earth had perhaps three hours of sunlight and three hours of night, and the moon loomed huge in the sky, at least during infrequent days when the impact-generated dust cleared. If the moon formed at about 2.4 Earth-radii from Earth's center (the same relative position as the outer edges of Saturn's rings), it would have subtended about 22° in the sky, looking forty-four times bigger than the moon looks to us! Probably, however, it was no more than a dull glow obscured by heavy clouds of dust and moisture during this period.

The tidal bulges, which force the moon to move outward, were strongest when the moon was closer to Earth. The moon therefore moved most rapidly outward at the beginning. By 4400 My ago, it was probably close to halfway to its present position, the month would have lasted nearly ten modern days (i.e., 240 hours), and Earth's rotation would have lasted about ten hours—five hours of dim, cloudy light and five hours of dark night. The day has continued to lengthen, as we will remind ourselves in coming chapters.

DOES EARTH'S MANTLE STILL SHOW TRACES OF THE GIANT IMPACT?

To geophysicists, one of the most interesting predictions of the Cameron-Melosh calculations is that Earth's mantle would have heated to temperatures well above the melting point. In current work, many investigators are looking for clues as to whether or not the mantle shows traces of this event. Almost all researchers agree that it must have been heated at least enough to melt. This conclusion comes from various lines of reasoning: (1) formation of the iron core seems to require at least partial melting; (2) theories of Earth's accretion require enough energy impacts to melt the upper mantle; (3) abundances of certain elements, such as of magnesium relative to silicon, could have been produced as molten material solidified and elements separated in different proportions in various types of solid mineral crystals. But the giant-impact theory predicts even higher temperatures. Does mantle chemistry really show traces of the very high-temperature melting calculated to have been caused by the moon-forming giant impact?

University of Arizona geochemists Michael Drake, his student Lisa McFarlane, and their colleagues have attacked this question by studying how

Hartmann

4450 My ago. As seen from the primeval moon, Earth was a dramatic sight because the moon was initially much closer to Earth than it is today. Here Earth fills much of the lunar night sky soon after sunset. Lava flows are creating the earliest lunar volcanic plains, destined to be pocked by impact craters. A reflection of the full moon can be seen in the center of Earth's night-side disk.

4500 My

4400 My ago. A view of Earth from the moon when the moon had receded to half its present distance. The lunar magma ocean has cooled, but the early lunar crust is broken by pools of erupted lava.

the mantle would have cooled and crystallized into solid minerals if it started out molten. Drake's group made laboratory measurements of mineral grain chemistry to study how compositions in the upper mantle would evolve as it cooled and solidified. At each temperature, as the cooling occurred, freshly solidifying mineral grains would interact with the remaining molten material. If the mantle had been completely molten, as predicted by the giant-impact theory, and if it had then cooled without too much stirring, the various elements would have separated into minerals according to their chemical affinities. Drake's group, as well as Australian researcher A. E. Ringwood, found discrepancies between the chemistry predicted by this scenario and that actually observed in upper mantle rocks. Unexpectedly, they did not find signs of the high-temperature melting period, and hence no clear proof of the moon-forming impact! They were forced to conclude that *either* Earth was never fully molten or, if it was molten, the rock crystals forming in the molten mantle did not interact chemically as much with their surrounding molten material as had been expected. The first possibility would mean either that the giant impact never occurred, or that the calculations overestimate the amount of heating of the mantle by the giant impact. The second possibility may be more likely, but it is complicated by our uncertainty about true mantle composition. For example, some geochemists suspect that the upper mantle has been relatively isolated from the lower mantle, so that upper mantle materials may not tell us much about lower mantle materials. As remarked in the last chapter, geochemists are still arguing about the nature of the mantle because the measurements involve state-of-the-art techniques. The uncertainties are frustrating to all would-be readers of Earth's story.

WHAT CHROMIUM TELLS US

One example of the incredibly detailed level of current studies of mantle history is work by Ringwood and his Australian colleagues on chromium (abbreviated Cr by chemists), vanadium (V), and manganese (Mn). Under conditions near Earth's surface, these three metals should show only limited chemical affinity for iron, so they should not have drained into the core with the iron. Yet their abundance patterns are different than expected had they stayed in the mantle. Meteorite rocks, including certain rare meteorites believed to have originated on Mars,* do not show this depletion pattern, as pointed out by University of Arizona geochemist Michael Drake and co-workers in 1989. The pattern is apparently unique to Earth's material. However, the moon does show this depletion pattern, and Ringwood cites this as evidence that the moon formed from Earth's mantle material. This chemical test, for a change, does support the giant-impact theory.

But why are the Cr-V-Mn abundance patterns different than expected in Earth's mantle? Did some of this material go into the iron core, or not? Ringwood was able to show that in a body as big as Earth, high pressures coupled with oxygen content in the innermost planet would act chemically to drive extra chromium, vanadium, and manganese out of the mantle and into the metal core.

Interestingly enough, Ringwood's work also indicated that in a planet as small as Mars, this process would not work, explaining why the Martian rocks still have their normal Cr-V-Mn abundances. And there was an unexpected bonus to this result. The Mars-sized impactor that blew off the Earth mantle material to make the moon would not show the Cr-V-Mn depletion, according to Ringwood's results. The fact that the moon does show the depletion proves, in Ringwood's view, that the moon's material had to come from the mantle of Earth, not from the mantle of the impactor itself.

*About a dozen meteorites are 1300-My-old fragments of basalt lava that contain gases matching the composition of Mars' atmosphere. Scientists believe they are rocks that were blown off Mars by impacts of moderately large asteroids and that later crashed into Earth. In the 1970s, no one believed such an event could happen, but research in the eighties convinced skeptics that these obscure rocks are actually from Mars!

TOWARD FUTURE WORK

Discussions between the geochemists and planetary physicists, and even among the geochemists themselves, have been vigorous since 1988. Clearly, the story of Earth's early history will come from combining many kinds of data. But for the time being, scientists from different disciplines tend to work in isolation and argue at conferences. At one meeting NASA researcher John Jones made the tongue-in-cheek gibe that "physics only shows what might have happened, whereas chemistry shows what really did happen." Yet the geochemical sessions of several conferences have featured arguments among the geochemists themselves, who because of the delicacy of the measurements sometimes find that they have conflicting results. Only a return to the lab can resolve such problems.

Perhaps the forthcoming geochemical measurements made by the next generation will clarify details of the moon's origin and the mantle's evolution. For now, the indications are that Earth suffered one of the most violent catastrophes of solar system history while giving birth to our satellite.

BOOK THREE

First Days on the Surface of a New World

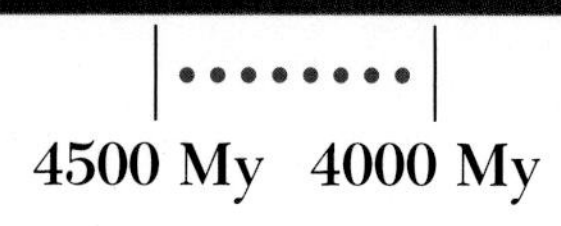

4500 to 4000 My Ago

6. The Era of Early Intense Bombardment

With the passing of each year, evidence is accumulating to indicate that the earth's cosmic relations may have had a great deal to do with its internal and crustal adjustments.

—*H. H. Nininger, pioneering meteoriticist,* Out of the Sky, *1952*

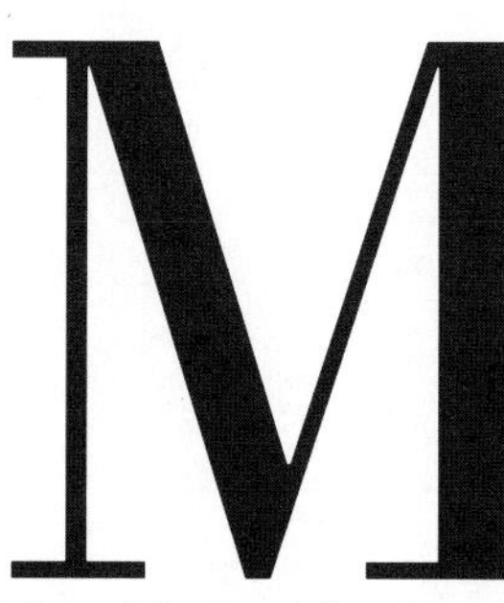

Most of the clues about the earliest events on Earth's surface have been destroyed by Earth's very active erosive processes. Although Earth is 4550 My old, rocks older than 3500 My are very rare. The oldest rocks that have been found date from about 3960 to 3500 My ago, a number of them from Canada and Greenland. The oldest reported examples, from 3960 My ago, were found in 1989, in the Northwest Territories of Canada.

Fortunately the moon comes to our rescue, helping us fill in the nearly blank pages of Earth's biography before 3500 My ago. The moon has had much less geologic activity than Earth, and hence older rocks are still preserved. The dark lava plains that make up the features of the man in the moon are about 3500 My old. The bright uplands of the moon, which cover roughly 90 percent of its surface, are even older, dating from before 4000 My ago. As mentioned earlier, astronauts could not find many lunar rocks older than 4100 My because they seem to have been pulverized by the intense cratering, but nevertheless most lunar rocks are in just the age range that is mostly missing on Earth. In short, the moon gives us a good opportunity to find clues about the earliest planetary conditions.

EARTH AND MOON— A STUDY IN CONTRASTS

Earth is four times bigger than the moon in size, and has six times as much gravity at its surface. These are important facts in understanding why Earth and its offspring are so different in terms of geologic activity and atmosphere.

Being more massive, Earth retained its heat more effectively. Heat generated in the center of the moon has only a fourth as far to go to reach the surface and radiate into space as does heat generated in Earth's center. Because it has remained hotter, Earth can still generate the active volcanoes and tectonic forces that tear continents apart. All these forces are dead on the moon.

The moon lacks atmosphere for two reasons. First, the material in the moon was heated to very high temperatures as it was blasted off Earth, so water and atmospheric gases were all driven off into space. The moon accreted dry. Second, even if you magically put an atmosphere on the moon today, the gas molecules would rapidly escape from the moon's gravity into space, because the moon's gravity is so weak. For these reasons, the moon undergoes no erosion by atmospheric agents: no rains disturb its dust, no glaciers slide down its mountains, no rivers cut canyons.

In the absence of such erosion, the moon retains impact craters—the scars left by the earliest planetesimals crashing into its surface. Earth's surface, in contrast, is geologically active. The average landforms on Earth are only about a tenth as old as those on the moon, and correspondingly fewer craters have accumulated.

THE EARLY INTENSE BOMBARDMENT CONTINUES

In contrast to today's Earth, primordial Earth was intensely bombarded because it was sweeping up the planetesimals left over from planet formation. How long did this sweep-up last? How did the bombardment rate change with time, and what were its effects on Earth?

The moon offers many of the clues we need. Because the moon has been alongside Earth since it formed roughly 4500 My ago, it must have experienced the same cosmic environment as Earth. It experienced the same bombardment by leftover planetesimals.

The bombardment history of the *Earth-moon system* as a whole can thus be studied by combining lunar and terrestrial cratering data. These data show that the cratering rates must have averaged much higher in the first few hundred million years than in recent geologic time. That intense cratering is called the *early intense bombardment.*

Detailed studies of the lunar craters reveal the magnitude of the early intense bombardment. The nine Apollo and Luna landing sites range in age from about 4000 to about 3200 My. Arranging them in sequence, we can count the number of craters per unit area at each site and deduce something about the rate of cratering throughout this interval time. The results show that the cratering rate 4000 My ago was as much as 1000 times higher than it has been since 3200 My ago, and that it was even higher before 4000 My ago.

How high did the rate go before 4000 My ago? To accrete Earth's entire mass in 70 million years requires something on the order of 1000 million times the present rate! Thus from 4500 to 4400 My

4400 My ago. Interplanetary debris left over from the formation of the planets continues to blast the clouded atmosphere of Earth.

ago, the cratering rate may have been hundreds of millions of times its present rate.

The young surface of Earth today has fewer craters than the moon, but the stable central plains of several continents are old enough to preserve remnants of ancient, eroded craters. Amazingly, this was not recognized until the 1960s, when abundant aerial mapping photos became widely available. Many ghostly circular patterns in landscapes on various continents began to be noticed by geological researchers. This was especially true in the ancient plains of central Canada, where many surfaces are 1000 to 2000 My old. In aerial photos of these regions, analysts found glacier-eroded remnants of many ancient craters, often taking the form of circular lakes. Canadian researchers such as C. S. Beals, Michael Dence, and Richard Grieve proved that these were impact scars and then were able to use them to refine our estimate of the average impact cratering rate on Earth in the last 1000 My. They find that it averaged about the same as the lunar rate in the last 3200 My.

From these quantitative clues, we can assemble a consistent overall picture of Earth's bombardment history. During Earth's accretion, it was 1000 million times the present rate, but by 4000 My ago it had already declined to 1000 times the present rate, and by 3200 My ago it had declined even further, roughly equaling the present rate. It has stayed roughly constant ever since.

The rate of decline measured from the craters matches exactly with the rate predicted from calculations of how fast Earth would have swept up the interplanetary planetesimal debris as it formed. These orbital calculations have been made by dynamical theorists such as George Wetherill, using computer models of the aggregation of Earth. According to Wetherill's results, the collision rate dropped by half in the first 20 My or so after Earth formed, and kept dropping as more and more planetesimals were swept up by the insatiable planet. The planetesimals with Earth-approaching orbits were swept up first, but the ones on more distant or more inclined orbits had to wait for the inexorable cumulative effects of gravitational disturbances to bring their orbits into configurations where they too could collide with Earth. Thus there was a continual slowdown in the rate of bombardment. The rate has leveled out in the last 3200 My because a small, steady supply of asteroid fragments and comets are perturbed onto Earth-approaching orbits at a constant rate from the asteroid belt and from the comet swarm in the outermost solar system.

EARLY BOMBARDMENT—AN ALTERNATIVE VIEW

When geochemists first studied lunar rocks and found out that few of them were older than 4000 My, they concluded that a sudden, short episode of cratering happened 4000 My ago, destroying all the older rocks. This was called the cratering cataclysm. One way this could have happened would be that another giant impactor hit the planet Venus 4000 My ago, throwing out a swarm of debris that hit Earth and the moon at that time.

The idea of a sudden, short cataclysm of cratering is clearly different from the slow-decline scenario described above. It is yet another example of the old conflict between catastrophist models and more uniformitarian models.

In this case, the cataclysm model lost favor by 1980, but was recently revived by NASA researcher Graham Ryder. Ryder notes that large impacts melt some rocks and produce some lavas, called impact melts. Thus, he says, the higher the impact rate, the more impact melts there should be. But it is very hard to find impact melts from before 4000 My ago. Therefore, Ryder proposed that there was very little cratering from 4500 to 4000 My ago, and then a sudden burst of cratering happened about 4000 or 3900 My ago, producing most of the lunar craters and lunar rock, and reshaping Earth's surface.

Many researchers, myself included, find this model harder to accept. We think that primitive rocks and lavas from before 4000 My ago are rare simply because they were pulverized by the intense early bombardment throughout the first 500 My. Furthermore, this fits the orbital calculations and cratering evidence. For example, as we have stated, to accrete Earth in the measured 70 My formation interval requires a dense interplanetary population of planetesimals. Orbital calculations indicate that it should have taken about 500 My to sweep up these planetesimals, and that the impact rate would have declined throughout this time. It is therefore hard to see how early Earth and the moon could have experienced 500 My with almost no cratering, followed by a burst of cratering.

In summary, there is no question that Earth was heavily bombarded at the beginning when it was formed some 4500 My ago, and there is no question that as late as 4000 My ago the bombardment rate was still hundreds of times higher than today, but the details of how this bombardment declined during the first 500 My still need to worked out.

The total amount of meteoritic material falling on Earth in the current geologic era, based on several different measurements, amounts to about 60,000 tons per year. That's 150 tons per day! Most of it is in the form of dust and small particles that go unnoticed, though many rock-sized meteorites fall

Hartmann photo

Earth's surface is so constantly reshaped by erosion that few meteorite impact craters survive. This impact crater, nearly a mile across, is an unusual exception, formed about 20,000 years ago. Meteor Crater, Arizona.

4400 My ago. Early Earth's surface. Undersides of clouds may have glowed red from countless volcanic eruptions as scattered areas of rocky crust began to form.

to the ground every year. The figures quoted above mean that before 4000 My ago, more than 60 million tons of material fell on Earth each day, including many large bodies. Gaping craters, from 100 meters across to many kilometers across, were being blasted out every year.

The important thing about all this is that the environment of Earth's surface during the first few hundred million years was entirely different from its environment today. It was kept in constant turmoil by a steady rain of incoming small meteorites and the sporadic impact explosions of larger, asteroid-sized projectiles.

ALIEN EARTHS

One of the most interesting features of our new "cosmic" view of early Earth is how much the planet has changed through geologic time. Most of us grew up with a half-formed idea that Earth's surface environment has been unchanging, except for the different kinds of animals and plants that lived at different times. This is the old Shakespearean "world's a stage" idea. The actors come and go, but the stage remains the same. In other words, according to this old, seemingly plausible view, if we could fly back through time in some sort of combination time machine/spaceship, and if we approached Earth at different times, orbited it, or made intrepid landings, we would find an atmosphere and geologic features similar to those of today.

Studies in recent decades show that this idea is untrue. Earth has been a succession of different, alien worlds.

We have already seen one example. For instance, in the first hundred My or so—Earth's infancy—the meteoritic impact rate was so high that the surface was melted and the atmosphere was shrouded in dust and often opaque. Some 4400 My ago, the surface would have been gloomy, glowing with the splashes and eruptions of the red-hot magma ocean, covered by only a thin crust of rock under a dark, glowering sky.

LOSS OF THE PRIMORDIAL ATMOSPHERE

The intense early bombardment also had an effect on Earth's atmosphere. The largest impacts tended to blast the original atmospheric gases into space.

What was the composition of the original, or so-called primary, atmosphere? We know that it could not have had as much oxygen as today's air, because today's oxygen is produced and maintained mostly by plants. There were no plants for the first 1000 My or so.

Instead of the familiar nitrogen/oxygen atmosphere of today, the atmosphere during Earth's initial growth was basically a concentration of the gases from interplanetary space, pulled toward Earth by Earth's gravity. Such an initial atmosphere would have been rich in hydrogen, the most common interplanetary gas and the same gas that forms most of the sun. Jupiter, with its enormous gravity, retains such an atmosphere today. (There was no danger of the hydrogen exploding, for there was no oxygen to allow it to burn.)

A hydrogen-rich atmosphere could not have lasted very long on Earth for two reasons. First, hydrogen atoms are so light and fast-moving that they rise to the top of the atmosphere and leak off into space. Second, and more important, as mentioned above, the giant impacts of the era of early intense

4300 My ago. Thick clouds of the early atmosphere block the sunlight and darken Earth's surface. Volcanic islands form amidst the seas of night.

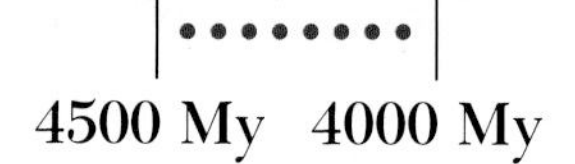

bombardment would have blown away the primordial gases. In particular, the giant impact that formed the moon probably blew away the first-generation atmosphere. Researchers Willi Benz, A. G. W. Cameron, and H. J. Melosh reported that according to their computer models of the giant impact, not only the atmosphere but much of the rock on one whole hemisphere was blown into space. Interestingly, since much of the rock was vaporized into a gas at temperatures of thousands of degrees, Earth temporarily had an "atmosphere" of hot silicon gases, as well as gases of iron, magnesium, carbon, and other rock-forming elements. That temporary "atmosphere" immediately recondensed into terrestrial rock material, but certain of the earliest gases could not recondense and were blown away forever.

VOLCANIC BUILDUP OF A "SECONDARY" ATMOSPHERE

Whence came the friendly atmosphere that surrounds our planet, shields us from raw sunlight, provides soft rains, and allows us to breathe? Part of the raw material for the present atmosphere came from the vaporous outpourings of primeval volcanoes. The volcanic gases were rich in water vapor and carbon dioxide, liberated from hot materials inside Earth. Even today, gases emitted by deep-seated volcanoes are still nearly 60 percent water vapor by weight, and about a fourth carbon dioxide, with smaller amounts of sulfur and nitrogen.* Primordial volcanic gases may have been even richer in water vapor. Thus the archaeologists of the atmosphere believe that a dense atmosphere of steam and carbon dioxide was established by about the time Earth reached its present size.

The evolution of this strange early atmosphere has been studied especially by Japanese researchers, in-

*One must be careful about this measurement. Volcanoes with shallow roots, drawing magma from the crust, simply recycle groundwater and other materials from the surface layers. The gases from volcanoes with deeper roots, reaching to the upper mantle, are more indicative of the original deep-seated raw materials emitted into early Earth's atmosphere.

4300 My ago. Lava flows pouring into the ocean begin to expand the realm of the land.

Miller

4200 My ago. The skies begin to brighten. Rains have removed the water vapor from the earlier thick cloud layer, building up the ocean and clearing the skies. Carbon dioxide and steam from lava flows continue to be major gases added to the atmosphere.

cluding Takafumi Matsui,* Y. Abe, and their colleagues. Starting with the usual ideas of Earth's accretion from numerous small planetesimals, they found that the cooling rate of the magma ocean and the general environment on primordial Earth were strongly dependent on how efficiently Earth retained the heat generated by the explosive impacts of the incoming planetesimals. This depended on the size distribution of the planetesimals, the mechanics of the impact process, and the duration of Earth's formative processes. Assuming Earth accumulated in 50 to 70 My, the magma ocean remained molten until Earth reached about 98 percent of its current diameter. This kept the surface and air too hot for much condensation of water or any formation of oceans until Earth finished forming about 4500 to 4450 My ago.

*I was struck by the international character of this scientific field when I heard Matsui lecture about his results at a conference in the Vernadsky Institute in Moscow. The Soviet audience was at the mercy of double translation from English through two non-native English speakers. Matsui spoke in Japanese-accented English, while a Soviet translator, fluent in English, struggled to understand Matsui's Japanese accent, and render the speech into Russian.

Throughout that time, Earth was pouring volcanic fumes of carbon dioxide and water vapor into the hot air at a prodigious rate. The air pressure and humidity rose. Matsui estimated that the early atmosphere may have been as much as seventy times as dense as now, approaching the oppressive carbon dioxide atmosphere that still exists on Venus, with its air pressure ninety times that on today's Earth. This general idea is supported by the calculations of Ann Vicary in the United States. She finds that the early cratering rates kept heating the surface rocks, driving off water and other volatiles, and maintaining an air pressure at least five times that experienced today.

ROCKS AND OCEANS AT LAST—4400(?) MY AGO

If you observe any modern volcanic lava flow, you can see that a thin scum of solid rock, inches thick, forms within hours. Within a few days, it is thick enough to walk on. Similarly, on the magma oceans of primordial Earth, such a thin crust of rock tried to form but was continually pulverized by new impacts. Thus no real solid surface landscape could form for millions of years.

As the impact rate declined, the heating from the impact explosions also declined. As a result of the decline in impacts and in heating, the magma ocean was less constantly stirred, and it also began to cool and solidify. Eventually a substantial solid crust took shape, first in patches here and there and then eventually over much of the surface.

Interestingly enough, we are not sure exactly when this happened. Certainly some sort of crust must have started forming by 4300 My ago; from that time come Earth's oldest known solid mineral crystals, some zircons found in Australia. Chemical studies of the moon's crust indicate it existed by 4400 My ago. Earth's crust had probably begun to form by the same time, consisting primarily of light-toned low-density feldspar-rich rock like the moon's highlands. Today's terrestrial crust has evolved far beyond that simple stage, as we will see in later chapters.

The first rock surface was barely exposed before it was covered with water. As long as the surface had been molten, all of Earth's near-surface water boiled and produced atmospheric water vapor. The atmosphere, in addition to being dusty-dark, was steamy. But as soon as the surface began to cool, the steam began to condense into liquid droplets, so that solid rock layers were soon swept by rains and dotted by lakes. Due to torrential rains from the cooling water-vapor atmosphere, more and more water accumulated in the lowest parts of the new lava surface. In this earliest competition between sea and land, the land was soon overwhelmed by a global ocean.

Partly on the basis of the Japanese studies, we can imagine a view of early Earth's surface after the closing stages of planet formation, starting around 4400 My ago. Not only was the air thicker and the pressure higher than today, but the primordial surface was blanketed by a high-altitude dust pall kicked up by the frequent impacts. The Japanese researchers picture a deep global ocean at this stage, but crater rims and volcanic islands may have risen above the waves here and there and red glows from scattered magma eruptions may have illuminated the undersides of the dark gray clouds. Neither sunlight nor the bright, impassive glow of the nearby newly formed moon could penetrate the dust shroud or cheer the scene.

One measure of the strangeness of Earth's atmosphere at that time is that we could not have lived in it because of the high pressure and lack of oxygen. Nonetheless, as the water vapor condensed out of the air to form oceans, the air pressure dropped toward values more familiar today.

TOWARD THE FIRST LANDSCAPES—4200 MY AGO

By 4200 My ago, the sky may have brightened somewhat. Deep, warm global oceans still covered much of Earth, under leaden skies and heavy tropical rains. The scene may have looked like a foggy, rainy evening in the Straits of Borneo. Steam clouds from undersea volcanic eruptions rumbled into the sky.

The cratering rate was still so high that the dominant large-scale landforms would not have been familiar continental masses or mountain chains, but ring-shaped mountainous rims of giant impact craters. In some places, these formed complete ring-shaped islands with enclosed central seas. In other places, arcs of individual islands, like beads on a string, traced the highest points of undersea crater rims. Erosion was fierce. Tides surged higher than today because the moon was closer than it is today. Massive tsunami waves occasionally swept the oceans' surface as the last planetesimals of the intense bombardment crashed into the sea. The first brave beginnings of land surfaces were fated to erode quickly under the relentless pounding of waves raised by the strong tides of the nearby moon. In the second competition between sea and

4200 My ago. The first moonrise. For most of the first few hundred My, Earth was probably too cloudy for the moon to be visible. After the clouds condensed into rain and formed the oceans, the skies began clearing intermittently in some regions. The moon had already receded from its initial proximity, but was still noticeably closer than today, as shown in this view of the full moon rising at sunset.

land, the earliest landmasses crumbled back into the water.

Clearly, this scenario indicates that the oceans are very old. In support of this, the structure of some of the oldest rocks gives testimony that the oceans were already well developed by at least 3300 My ago or earlier.

As large impacts blew away parts of the atmosphere, and as carbon dioxide from the atmosphere dissolved in the accumulating ocean waters, eventually converting to carbonate rocks, the atmosphere grew thinner and clearer. Between major impacts, the first sunny days may have occurred. Eventually the deep global ocean envisioned by Matsui began evaporating, exposing permanently higher-standing crustal regions as the bare land surfaces.

An important question is when true continental masses began to appear. True continents are not merely high spots of landmasses, sticking out above the waves. Continents are concentrations of low-density granitic rocks floating in the higher-density mantle rock the way wood blocks float in water.

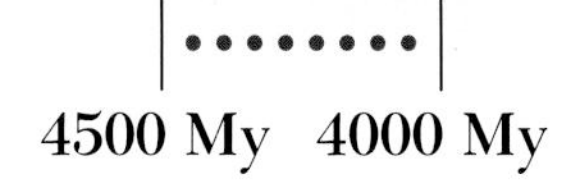

Steam clouds billow into the sky as molten lava pours from underground vents into the sea. Foreground lava is only a few weeks old. Kalapana, Hawaii.

True continents could not arise until geological processes created concentrations of low-density granitic rock. This did not happen all at once, but required some extensive reprocessing and differentiation of the solid rocks produced by the solidification of the magma ocean.

One way of beginning the buildup of continental landmasses involved giant impacts. While excavating the crust and forming great circular basins, they piled up masses of ejected materials in certain regions outside the basin rims. Random juxtaposition probably produced thick piles of crustal ejecta in some regions. This might have mimicked continents. Two big basins could have piled enough debris between them to create a broad, dry landmass.

Even after major landmasses began to appear, perhaps by 4300 My ago, the orbital crews and landing parties from our imaginary spaceship/time machine would have seen landscapes very different from those of today, because there were no plants or animals. Photographs might have revealed scenes such as those on high mountain peaks or deserts today. No flowers or grasses graced the plains. No trees cast shade. No foliage rustled and no bird chirps or mating calls broke the sound of the wind hissing across the rocks. The major landforms were still circular craters, lava flows, and fractures, like those mapped on Venus by the 1990 Magellan spacecraft.

AN ALIEN AND UNSTABLE ENVIRONMENT

We have mentioned that the moon formed close to Earth, and that in the early eons it was closer to Earth than it is today and raised higher tides. However, most of its outward motion occurred within the first few hundred million years. By 100 My after it formed—say, at about 4400 My ago—the moon was already halfway to its present distance. Earth's rotation period had lengthened from around five hours to around ten hours, still a strangely truncated day/night cycle compared with our own twenty-four hours.

Even the climate on this uninhabited world was strikingly alien.

During the early intense bombardment, the climate must have been unstable. Air pressure may have suddenly dropped when a large impact blew part of the atmosphere away; then air pressure would climb as volcanic gases accumulated. Temperatures may have fluctuated wildly, depending on the amount of CO_2 and the amount of sunlight getting through the impact-generated dust palls.

Impacts of serious effect happened sporadically until about 3800 My ago. Dates of lunar rocks suggest that at least one or two 1000-kilometer-wide basins were formed on the moon as recently as 4000 to 3800 My ago. Considering the larger area and gravity of Earth, our planet probably received several such giant impacts in that time. You might think that the explosion would affect only one side of Earth. However, University of Arizona geophysicist Jay Melosh has shown that debris would shoot out and fall back into the atmosphere on all sides of Earth, causing a tremendous heating. According to calculations by Melosh, NASA geophysicist Kevin Zahnle, and others, such catastrophic impacts dumped enough energy into the atmosphere to vaporize the fledgling oceans. Thus these scientists believe that even as "recently" as 3800 My ago, the oceans may now and again have been vaporized, only to re-form in decades as the air cooled and massive rainstorms raged across the globe.

Intense impact conditions guaranteed rapid, unpredictable changes in climate. Just when clearer, more familiar conditions began to develop, a new large impact might boil off most of the oceans and spread a pall of dust around the world, ushering in a decade of night. Such effects were studied in the late 1980s by Carl Sagan and his colleague David Grinspoon, who proposed wild climatic oscillations.

The profound environmental effects of Earth's early intense bombardment from 4500 to 4000 My ago are just beginning to be recognized by researchers in geology, geochemistry, and biology. Even by 4000 My ago, as the bombardment was dropping toward today's rates, conditions were strange. The moon had moved out to about 70 percent of its present distance, giving a day/night cycle of thirteen and a half hours. The atmosphere remained mostly carbon dioxide (CO_2). As we will see in the next two chapters, the climatic instability may also have had the important side effect of delaying the evolution of life for several hundred My. As a result of the new interest in this mysterious period, we can expect more vigorous research about Earth's primordial environment in the coming decade.

4200 My 3500 My

4200 to 3500 My Ago

7. The Most Important Event in the Solar System

> Life—the word is so easy to understand, yet so enigmatic for any thoughtful person.
>
> —*A. I. Oparin,* Life: Its Nature, Origin, and Development, *1962*

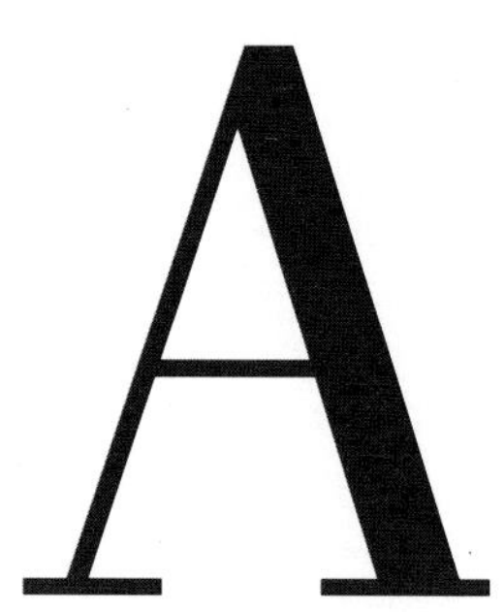

t first glance, the long years from 4200 to 3500 My ago look like the most uneventful of the planet's history. The last remaining interplanetary planetesimals were being used up by collisions with planets, and so the rate of sporadic impacts continued to drop. Earth was coming out of the tail end of the intense bombardment by 3800 My ago, and by 3400 My ago the rate was similar to that of the present day, and impacts no longer seemed an important planet-shaping force. Broad landmasses had appeared as ocean

4200 My ago. The dense, moist, CO_2-rich atmosphere of early Earth maintained a cloudy, hazy environment over much of the surface even as the skies continued to lighten. Oceans have formed from water vapor condensed into rain; early landmasses were primarily mountain ramparts piled up in association with impact craters.

water evaporated into the thinning atmosphere. Little biological activity enlivened the seas and virtually none graced the fledgling continents. Earth was barren; geological processes were painfully slow.

If our imaginary time traveler could have visited Earth during this era, each day, each million years, might have seemed like an endless waiting. The sun rose and set over endless waves and barren mountains. Land surfaces eroded with abandon; soil was carried into the sea. The moon, already at most of its present distance by 4 billion years ago, continued to move away very slowly. Winds blew, rains came and went. Lightning flashed over scenes where no beings were present to hear the thunder.

To a casual observer, there might have seemed to be nothing new under the sun.

But paradoxically, these seemingly quiet days witnessed the most startling event in solar system history—the event that more than any other makes Earth unique. It was not the mighty smashups of planetesimals to make new planets or the catastrophic collision that made the moon. Earth was in its adolescence, and as is the case with all adolescents, the most magical things were happening slowly and out of sight. Deep in the seas, chemical processes were producing life.

WHAT IS LIFE?

The origin of life on Earth has been one of the most fascinating and frustrating scientific mysteries of all time. We can talk about it only if we answer another question first. What exactly do we mean by life? This is a tricky question, because our sensations of life do not guide us accurately toward describing its physical and chemical processes. To answer according to the scientific method of hypothesis and experiment, we have to start with physics and chemistry.

During most of history, leading thinkers would have dealt with the question in a more subjective, introverted way and would have given answers wildly different from those of the late twentieth century. In the earliest writings, for example, life was thought of as a divine property instilled by the gods. In the Victorian era, life was often viewed mechanistically, as a manifestation of an undetectable energy field, sometimes called the "vital force"; this view was called vitalism. Shortly after the discovery of electricity, scientists were impressed that electrical impulses could trigger muscle reactions; some researchers speculated that electricity was the vital force. This is why the makers of the 1931 film *Frankenstein* had the mad doctor bring his creature to life by exposing it to lightning during an electrical storm.*

Modern scientific studies of life's origin take nothing away from its unique and holy quality. Rather, they are aimed at understanding the processes that lead to life. Through this approach, we have learned that life is best viewed as a *process*, not a *condition*. The stopping of the *process* is what we call death.

An important feature of life is that it involves taking in material, chemically processing it, adding some of it to the material of the creature, and eliminating the waste. As early as 1962 the Soviet scientist A. I. Oparin made an interesting analogy. He pointed out that some people had likened a creature such as the human organism to a bucket of water, in the sense that the skin encloses a fluid medium. But Oparin noted that a better analogy would be a bucket of water with a spigot at the bottom. The bucket is placed under a constantly flowing faucet; water enters the top of the bucket and continually drains out the bottom at the same rate. Even though we might still call this a bucket of water, it is not a fixed thing. It is now a dynamic system, and the water is never the same at any two given instants. This analogue emphasizes that a living creature is constantly changing.

Another analogue is a candle flame, which draws its sustenance from the candle's wax, oxidizing it and transforming it into ionized gas (the flame), and exhausting the waste products in invisible gases and thin smoke. The wavering, dancing flame looks like a well-defined entity, and is given the name *flame*, as if it were a fixed *thing*. But in reality, its physical makeup is constantly changing, and it never consists of the same atoms from one second to the next. A wisp of smoke from the flame is a thing in the sense that all its atoms could be labeled and they all stay the same as the wisp moves across the room. But the atoms in the flame are not constant. Its shape and color may look constant, but it is more a process than a thing because its atoms constantly change. Our language fails to make this distinction and hence misleads us.

Similarly, the cells in our body are constantly replaced, and today we have few of the cells we started out with. Life is a process. A living organism is not a fixed thing, but a system that is processing material.

Of course, another critical quality of life as we understand it is its ability to reproduce itself. This

*Actually, there is no such scene in Mary Shelley's original 1818 novel. Victor Frankenstein said he would not give away the secret of how his monster was given life.

ability, a source of joy and desire and pain in ordinary life, holds keys to understanding what happened in the seas 4 billion years ago. It leads scientists on a quest toward the smaller and smaller—a quest to understand life processes on a molecular level. Life's ability to reproduce itself involves some extraordinary qualities of carbon-based molecules.* These properties can be studied only in the context of the submicroscopic world of the atom and the molecule, but they help us to understand both the nature and the origin of life around us.

Carbon's special ability is twofold. First, carbon atoms have a strong ability to bond to each other and to atoms of hydrogen, oxygen, nitrogen, and other elements to make very big molecules containing thousands of atoms. These molecules are much bigger than common molecules. The common molecule, water (H_2O), for example, is a mere three-atom molecule with two atoms of hydrogen bonded to one of oxygen. In contrast, one of the simplest carbon-based molecules, methane (CH_4), contains five atoms and has the property of being able to link with more CH_3 and CH_2 units to make a whole series of compounds from ethane (C_2H_6, with eight atoms) to hectane ($C_{100}H_{202}$, with 302 atoms) and beyond. These are relatively simple carbon-based compounds. There is no hard-and-fast dividing line between simple and complex carbon-based molecules, but they can get much more complex than hectane. The giant protein molecules inside cells can contain a million atoms!

The second aspect of carbon is that certain big carbon-based molecules have the ability to split down the middle into two halves that are carbon copies, so to speak, of each other. This is the microscopic chemical secret to reproduction. It is the quality we will focus on in understanding how life originated.

*All life as we know it is based on the unusual abilities of carbon atoms to combine into giant molecules with peculiar properties. People often ask whether there might be life of totally unfamiliar kinds, perhaps based on the similar properties of silicon atoms or based on something quite different, such as the organization of magnetic fields. But we know of no other such lifeforms, and in this book we will limit ourselves to the kinds of life witnessed on our own planet.

The complex molecules based on carbon, especially those with carbon-to-carbon bonds, are called *organic molecules*. The study of their reactions is called *organic chemistry*. The adjective *organic* does not necessarily mean that the molecules are produced by living material or that they involve biological processes. Much of organic chemistry involves nonbiological reactions. For example, the understanding of polymers, which are nonbiological substances involving long chains of organic molecules, led to the development of the plastics industry after World War II. *Biochemistry* refers to the chemistry of molecules involved in life processes.

The quest for the origin of life thus becomes an attempt to understand how organic chemical processes built ever more complex molecules, and how these led to biochemical processes. Chemical research has not yet produced true living systems in the lab, but it has given us a crude idea of how living creatures evolved from these molecules on Earth, and why they did not evolve on other planets in our solar system. Here is our best estimate of that sequence of events.

COMPLEX MOLECULES

The first appearance of self-reproducing molecules in the long-lost oceans of ancient Earth was a key step, and the water of the early seas was the perfect medium for it. As we saw in the last chapter, the thick air above the oceans was rich in carbon dioxide. Carbon dioxide readily dissolves in water, as drinkers of carbonated beverages know. Thus, as soon as the warm primeval oceans appeared, they began to gain a concentration of carbon dioxide, assuring at least one source of carbon atoms.

What happened next can be visualized from a simple model. Imagine a molecule that is shaped like a necklace—a string of different-colored beads. Each color represents a different element: carbon

(C), hydrogen (H), oxygen (O), and nitrogen (N) were especially important ones. Now suppose that in this *particular* arrangement, each atom on the string has the possibility of joining with another atom of the same element. Imagine this necklace-like molecule immersed and drifting in the primordial ocean, which is a soup of C, H, O, and N atoms as well as water, carbon dioxide, and other molecules. According to the overly simplified properties we are imagining for this molecule, whenever one of its C atoms bumps against another C atom, the second is joined to the first. Likewise with the other elements. Thus what started out as a single string of beads builds itself into a double string of beads, side by side—each red bead glued to another red bead, a blue paired with each blue, and so on. The initial necklace is now a double parallel strand, like the two sides of a ladder. The whole thing might be straight, looped, or with the two strands twisted one around the other. In this imaginary model, it might look like this:

. . . CHONCHONCHONCHONCHON. . .
. . . CHONCHONCHONCHONCHON. . .

Now we must imagine another property that represents the actual behavior of carbon molecules. The bonds between the atoms *on* each strand are stronger than the bonds *between* the two strands. Thus under the influence of disturbances, the whole assembly may split back into two single strings. It is as if you glued two identical necklaces together bead by bead and then dropped the whole assembly on the floor. *Presto chango*: you have two identical single-strand necklaces again.

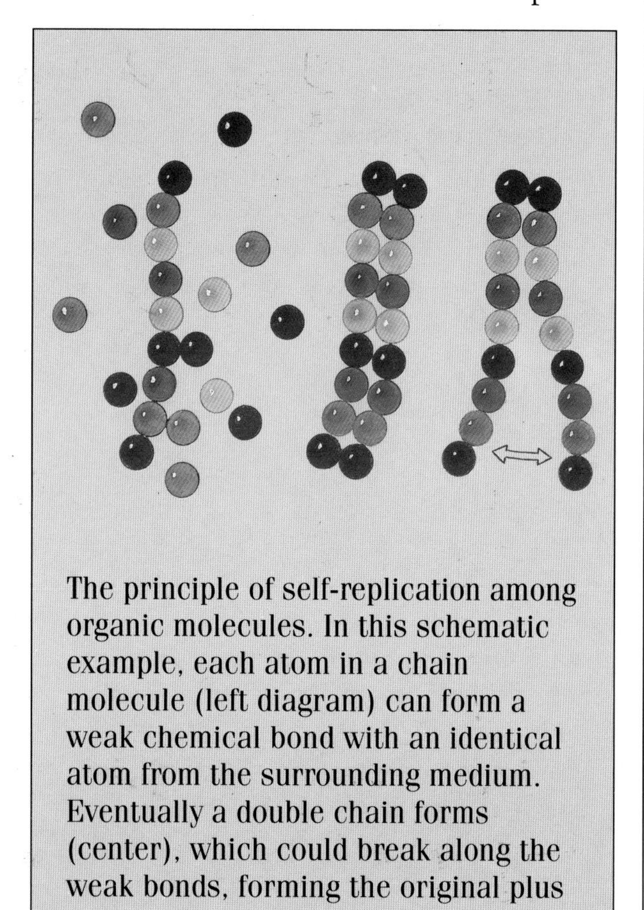

The principle of self-replication among organic molecules. In this schematic example, each atom in a chain molecule (left diagram) can form a weak chemical bond with an identical atom from the surrounding medium. Eventually a double chain forms (center), which could break along the weak bonds, forming the original plus a copy of the original (right diagram).

The astounding thing about this process is that the first molecule, purely by adding atoms according to its chemical properties and then breaking apart, *manufactured* a copy of itself. Now each of the single strands goes through the same scenario all over again. After many random collisions with various atoms, the first one adds atoms to itself again, and the second one, which is indistinguishable chemically from the first, does likewise. We now get *two* double-strand necklaces. To put it another way, the complicated double-strand necklace has reproduced itself. When they eventually break, we have four single-strand necklaces, which can make four double-strand copies. We will soon have eight, sixteen, and more copies.

It takes no genius to see that loose carbon, hydrogen, or other appropriate types of atoms in the seawater will eventually be swept up and concentrated in these big molecules. This is a crude description of the beginnings of life, through organic chemistry. The big molecules are not literally living matter, but we might figuratively say that they are "feeding" on the individual C, H, O, N, and other atoms that they find in the soup. The C, H, O, N, and fragmentary combinations of them might be regarded as "food" for the prospering, growing, and reproducing organic molecules. Due to the curious but purely

4100 My ago. Earth adds new land in the midst of the seas. Volcanic eruptions piled up lavas on the sea floor, forming undersea mountains. At some places, these broke through the surface waves, forming active volcanic islands. The same process continues today. The intense early bombardment was dwindling, but a contrail across the sky marks the passage of a meteorite that flashed overhead a few minutes before this view. In the surrounding seas, chemical reactions are building up the concentration of organic compounds.

chemical properties of organic molecules, what started as an ocean of simple water, carbon dioxide, and other atoms and molecules begins to acquire a high concentration of complicated organic molecules doing complicated things!

Our example is definitely oversimplified. For example, the elements in the first necklace would not have to attract another atom of the same element. Suppose we have a long chain,

...C H O N C H O N C H O N C H O N C H O N...

and instead of bonding with another C, the C atom bonds with an H atom; the H atom with an O; the O with an N; and so on. Then we would eventually get a double strand like this:

...C H O N C H O N C H O N C H O N C H O N...
...H O N C H O N C H O N C H O N C H O N C...

When it broke down the middle, the lower strand would be essentially a duplicate of the upper. A somewhat more realistic view of some biochemical processes goes like this: in the initial chain, the Cs might bond with other Cs, the Hs with some element X, the Os with some element Y, and so on. Then we could build up a chain like this:

...C H O N C H O N C H O N C H O N C H O N...
...C X Y Z C X Y Z C X Y Z C X Y Z C X Y Z...

Now imagine that this double chain breaks down the middle. The lower chain becomes a template for manufacturing another copy of the upper chain, because as it drifts off, the C will bond with a C, the X will bond with an H, and so on. Eventually it will build a new double chain:

...C X Y Z C X Y Z C X Y Z C X Y Z C X Y Z...
...C H O N C H O N C H O N C H O N C H O N...

When *that* double chain breaks apart, we see that the template will have manufactured a second copy of our original CHON chain. Thus we have molecules that create templates to reproduce themselves.[*]

[*]More detailed discussion of the chemical processes of life is beyond the scope of this book, but an unsurpassed introduction can be found in Richard Dawkins's book *The Selfish Gene.* Dawkins goes beyond a description of the origin process to describe how natural selection could create a diversity of different species of molecules, and eventually of creatures.

Many different complex organic molecules have properties similar to those described above. Imagine, for example, that one species of "growing" molecule includes only C, H, O, and N atoms, while a second species includes some phosphorus (P) atoms and fewer Ns. Imagine that the P atom concentration in the water was initially very low. Then the production rate of this second species was slow, because molecules would only rarely bump into P atoms. Suppose that eventually the first and second species together have consumed most of the Ns. Their production rates would slow down because few Ns would be available to build new molecules. Conceivably the P-bearing species might attract N more readily, perhaps by having a stronger bond with N. Now the second species might reproduce faster, even though it was initially rarer. Thus a simple ocean in which one organic molecule overwhelmingly dominated might evolve into an ocean with two types competing for the C, H, O, and N raw materials, and the second species might begin to take over from the first.

To take a more spooky example, suppose that a still larger molecule formed with these properties. Suppose that, by chance, one portion of its giant string of atom-beads was exactly like the strand that we discussed in our first example. Then, every time it encountered one of our first species, it could absorb the whole thing ready-made into itself. This third type would thus be "feeding" on the first type.

These are merely metaphors for the complex properties of these early molecules. One can imagine ever more interesting properties. For instance, a giant molecule that had a shape consisting of an "armor" of unreactive atoms on its outside, but a hollow center, could pass seawater through its interior, absorbing the atoms and molecules it needed for growth, yet shielded from being absorbed by a still larger molecule that it might encounter. Thus each structure and each particular chemical property might have its own survival value.

The fascinating thing about this story is that we are discussing merely random encounters and chance events, but the chemical properties lead to incredibly complex "behaviors" that mimic those of simple lifeforms.

WHAT THE TEST TUBE TELLS US

Are these simplistic tales of molecules really relevant to the origin of life? Is there any way to verify that complex molecules really built up in these ways? The answer to both questions is yes.

One of the most important types of complex organic molecules is the class known as amino acids. These are not lifeforms, but they are the essential building blocks for still bigger molecules called proteins, which in turn are essential ingredients of cells. In order of increasing complexity, these components are:

- *organic molecules*, which can join to make
- *amino acids*, which can join into long strings, making
- *peptides*, which can form long peptide chains that are
- *proteins*, which agglomerate with other material to form
- *protoplasm*, which is the fluid, complex medium in cells.

This hierarchy, from organic molecules to protoplasm, is not the whole story of forming life, because there are other important giant molecules we have not discussed. Among the most famous are two

3900 My ago. Origins of life. Simple single-celled and many-celled microorganisms were the earliest to evolve, "feeding" on chemicals in the seawater.

types of giant molecules found in the nuclei of cells. They are called nucleic acids. One is ribonucleic acid, or RNA, and the other is deoxyribonucleic acid, or DNA. A DNA molecule may involve roughly 100,000 atoms of C, H, O, N, P, and other elements, arranged in complicated strings, chains, and rings. The DNA and RNA molecules in the nuclei of our cells contain the sequences of atoms that "direct" the reproduction of cells.

While scientists have not gone so far as to create living material in the lab, they have shown how the amino acids grew from primeval atoms and simple molecules. This was first demonstrated in the laboratory in 1953, when University of Chicago graduate student Stanley Miller devised an experiment to simulate life-forming conditions in the early environment. Miller used a beaker containing steam, methane, ammonia, hydrogen (to represent the early air), and water (to represent the early oceans). Steam from the "early ocean" circulated through the "early air" and recondensed. To represent lightning from early thunderstorms, he passed sparks through the gaseous mixture. In a matter of days, the clear "ocean" in the beaker had turned muddy red. The dark color was from a thick concentration of amino acids, synthesized from the water and gases with the help of the energy from the "lightning."

This so-called Miller experiment was a breakthrough in understanding what happened on primordial Earth. It proved that the building blocks of life could be created on Earth from purely natural processes in natural conditions. The experiment was repeated in test tubes and beakers in many labs, with different "early atmosphere" compositions and different physical conditions. The amazing thing turned out to be the ease with which amino acids form. What had been thought to be a difficult step on the road to life occurred readily. Many different energy sources worked. Complex amino acids were formed even when impacts (representing meteorites) were substituted for lightning as the energy source!

According to these results, amino acids might form anywhere, not just on early Earth. Sure enough, scientists soon found that certain carbon-rich black meteorites, called carbonaceous chondrites, are also rich in amino acids. Microscopic study shows that liquid water permeated the cracks and pore spaces in these meteorites a few million years after they formed, 4500 My ago. Apparently, the parent planetesimals, whose fragments these meteorites are, originally contained ices or chemically bound water in their minerals. After mild heating, the water was released as liquid. Nature took its course, and a concentration of amino acid molecules accumulated. Searches of these meteorites revealed no trace of fossil lifeforms, and so the process apparently went no further in these primitive bodies.

On Earth, conditions were clearly more hospitable for further evolution.

HOW TO MAKE A CELL

Why haven't scientists created life in the lab? The simplest lifeforms are one-celled microscopic organisms. The basic unit of life is the cell, which is far more complicated than the individual amino acids, proteins, and other building blocks we've been discussing. The cell is generally defined as "the smallest unit of life that still retains the properties of life," to quote one leading biology text (Starr and Taggart, 1981). Generally, therefore, the goal in studying life's creation has been to learn how "the first cells emerged through the evolution of complex systems of molecules," to quote the same source. Most scientists approach this through studies of the behavior of complex systems of organic molecules.

It is not easy. If scientists leave the Miller experiment running for longer and longer intervals, for example, they merely get more amino acids, not cells. Presumably the early terrestrial environment was richer in complex compounds and processes than the scientists' test tubes.

Could chemical processes in the early oceans have produced anything that looked like a cell?

Here again, test-tube results are remarkable. As early as the 1930s, the Dutch chemist Bungenberg de Jong experimented with solutions of organic molecules. He demonstrated certain cases in which if organic molecules were introduced into a test tube of water, the molecules *spontaneously* agglomerated into microscopic spherules about the size and shape of cells. The spherules are called *coacervates.* In de Jong's experiment, what started out as an "ocean" of dispersed organic molecules ended up as cell-sized spherules floating in relatively clear water!

Still another example was found in the 1950s when University of Florida chemist S. W. Fox and his colleagues prepared solutions of big protein-like molecules, called *proteinoids,* in hot water. When Fox let them cool, he found that the proteinoid molecules glommed together into cell-sized

3900 My ago. A cratered landscape. Impact craters remained one of the most common landforms on the earliest continents. We view a one-kilometer-wide crater in profile on the coastline in the distance, as seen from among the rubbly boulders blown out by another crater near where we are standing.

3700 My ago. As water evaporated from tidal ponds, organic matter in the remaining pond water became more concentrated. This concentration may have helped promote evolution of colonies of algae-like scum composed of simple organisms.

spherules, and even the spherules tended to link together into chains.

So it is clear that, in nature, organic molecules tend to aggregate into complex cell-like systems in which further reactions can take place.

In a sense, this makes things sound too simple. We've gone a long way toward a cell, yet no one has produced living matter. However, new discoveries continue to be reported around the world. For example, a 1990 report by French and Swiss chemists in the *Journal of the American Chemical Society* discovered "behavior" similar to reproduction in tiny coacervate-like droplets suspended in an organic solvent. When certain molecules from the solvent interact with the surface layer of the droplet, lithium hydroxide in the droplet triggers a splitting of the droplets into more and more smaller, cell-like spheroids. The researchers say that this process, while still purely chemical, mimics cell reproduction better than other processes yet studied. The greatest current enigma is the "magic" step between these nonliving cell-like systems and truly living, reproducing cells.

I say "magic" not because we believe that the transition was supernatural (whatever that means), but merely because we still don't understand it. The missing link in the process may come from the somewhat controversial views of biologist Lynn Margulis, at the University of Massachusetts at Amherst. She views the cell not so much as an individual entity or first step in life, but as a *symbiotic population* of more primitive semi-organisms that prospered by aggregating communally into cells (Mann, 1991). Therefore, the emphasis in learning how cells evolved may be not so much in experimenting with evolution of complex inert molecules, to paraphrase the goal mentioned above, but to learn more about the symbiotic reactions *among* the groups of complex organics. Such reactions allow them to prosper when they aggregate in cell-like structures, where RNA and DNA regulate the reproduction of not only single molecules but also whole cell-sized units. Regardless of the seeming "magic" or subtlety of the interactions, most scientists believe that the whole process was still a process of chemistry that might have happened not only on Earth but even on other planets, if not in our solar system then near other stars.

DELAYING THE ORIGIN OF LIFE

Probably organic molecules were present on Earth since the planet's formation, and *prebiological* evolution of organic materials probably began as soon as there was a nonmolten surface with lakes and oceans of liquid water collecting on it. The molecular processes we have been considering—the prebiological aggregation into coacervates and other cell-sized droplets—may have been active at least 4200 My ago.

However, the oceans needed to be relatively stable to allow the molecular processes to produce simple lifeforms, and for the simple lifeforms to evolve into more complex lifeforms. But as we saw in the last chapter, the early intense bombardment may have delayed the stability of oceans. Until 4000 to 3800 My ago, the oceans were intermittently being vaporized by giant impacts. So, according to the ideas of Cal Tech planetary scientist David Stevenson, NASA researcher Kevin Zahnle, and others, the origin of life at more than a molecular level was frustrated by the impacts until perhaps 3900 My ago.

Zahnle points out that when the oceans were vaporized by giant impacts, the only environments in which organisms might survive were the sludges on deep-sea floors. According to this idea, the common ancestors of all species might have been bacteria that, by the chance of being sea-floor organisms, survived the last colossal impacts. Zahnle claims support for his theory by citing comparisons of DNA and other genetic material among different species. These comparisons suggest that our common ancestors were indeed heat-loving sulfur-metabolizing bacteria similar to those found on deep-sea floors today—"the kinds of creatures,"

4200 My 3500 My

3700 My ago. Our algae ancestors. As large numbers of early organisms evolved in the seawater, mats of floating algae scum might have been some of the first visible evidence of life on Earth.

Zahnle remarks, "that would have been quite happy" when Earth's atmosphere was heated and the oceans mostly vaporized by colossal impacts.

THE EARLIEST FOSSIL EVIDENCE OF LIFE

Sometime, perhaps by 3900 My ago, true cells emerged. This event is lost in the mists of time. Nonetheless, by 3500 My ago, enough cells existed to begin leaving fossil traces. The oldest microscopic-scale fossils, which seem to be remnants of clumps of cells, date from this era, about 3500 My ago. For example, strings of microscopic cell-like forms have been found in western Australia from about 3500 My ago, and cell-like spheroids and rod-shaped forms have been found in the so-called Fig Tree group of chert rocks from South Africa, and in sediments from nearby Rhodesia, dated at about 3200 to 3100 My ago. Incidentally, as we will see in Chapter 11, western Australia and southeastern Africa may have been a continuous landmass at that time, so we may be seeing the earliest lifeforms from what was then a single region. Interestingly, bubbles in the 3500-My-old Australian rocks contain gases that have been studied as possible samples of the actual ancient air! More recent microfossils from around 2000 My ago are found in North America, in the ancient rocks of central Canada, near Lake Superior.

Earth has been so geologically active that only a few traces can be found of rocks formed in this middle-age era of Earth's development. Rains erode them, colliding continental blocks crush them, volcanoes melt them, and sediments bury them under hundreds of yards of rock and soil. To make matters worse, the most ancient fossils represent only microscopic lifeforms that are difficult to recognize except in cases when they are massed together. As a result of these difficulties, the earliest fossils were recognized only in the last few decades. More discoveries can be expected in coming years.

THE CLIMATE WHEN LIFE WAS BORN

Not only the impact environment had an effect on early Earth. The relatively young sun was subtly different during this time: it was perhaps only 75 percent as bright as it is today. How different was Earth's climate due to this effect? Does this mean that Earth was much colder? If the sun of 3800 My ago was 75 percent as bright as today, the reduction in radiation alone would place the mean temperature at mid-latitudes not at today's room temperature but near freezing. This, however, is too low an estimate. Because the atmosphere was still composed of moist carbon dioxide, a strong greenhouse effect prevailed. Calculations of the early greenhouse effect during this era, summarized by geologist Preston Cloud, indicate mean temperatures around 11°C, or 52°F, still significantly cooler than today but warm enough to guarantee liquid oceans, which are also known to have existed from geological deposits.

In addition to cooler temperatures, Earth's shorter day and somewhat closer moon meant that tides continued to be more pronounced during the era of life's formation than today. This in turn helped produce large tidal pools of seawater along coastal inlands, where evaporation could remove the water and leave the organic soups more concentrated.

It was thus in restless seas and transient tide pools and under cooler, clearing skies that the first organisms evolved.

BOOK FOUR

An Unfamiliar Earth and a Midlife Crisis

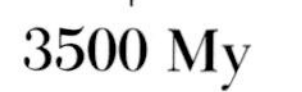

3500 My 2000 My

3500 to 2000 My Ago

8. Alien and Alternative Lifeforms on Earth

> Out of nothingness arises the sign of infinity; . . . The land and the water make numbers joined, a poem written with flesh and stronger than steel or granite. Through endless night the earth whirls toward a creation unknown.
>
> —*Henry Miller,* Tropic of Cancer, *1934*

How direct was the line of evolution that led from the first single-celled creatures to human beings? Textbooks sometimes show tidy evolutionary sequences with one species following the next, as if we were witnessing a circus parade down Earth's Main Street. Instead, there is a gnawing suspicion that evolution may have been quite indirect, with many false starts along now-abandoned evolutionary paths. A start on one path may have been canceled by some environmental or biological disaster, whereupon evolution was sent back to Go and had to start over. Some of the strangest recent paleontological discoveries suggest that Earth may have once hosted lifeforms that were weirdly different from our own ancestors.

CREATURES FROM SPACE?

The most radical idea of this sort is that the first lifeforms were truly alien in the sense that they, or at least their organic materials (amino acids, proteins, or nucleic acids), evolved somewhere other than Earth. This is not impossible since we have already found extraterrestrial amino acids in carbon-rich meteorites. Certain of these carbonaceous meteorites show clear evidence that liquid water percolated through them, probably from melting ice. The parent bodies of these meteorites were probably the black asteroids and comet nuclei that contain carbonaceous soils and ice. Organic chemical evolution is so easy—so fated—that these moist, carbon-rich worldlets quickly produced at least amino acids, if not more complex materials. We can be sure that many such asteroids and comets fell on primordial Earth as it formed, "seeding" it with prebiological material. Some researchers postulate that Earth received even more complicated materials, such as proteins, viruses, or even living organisms, that evolved inside early interplanetary bodies.

This idea has been around for several decades. In a way, it is not too important, because we know that amino acids would have been soon synthesized on primordial Earth, whether they arrived in meteorites or not. Miller's experiments showed that.

A new twist to these ideas, however, is that Earth could have been seeded with alien amino acids, proteins, viruses, or lifeforms that might have prospered except that they were all destroyed by the intense cratering of that era. In other words, the very intensity of the meteorite infall rate, which allowed these materials to accumulate rapidly, may also have destroyed them. The big danger came from the rare big impactors. As mentioned in the last chapter, the biggest impacts probably temporarily vaporized the oceans, thus wiping the slate clean of any products of alien evolution and allowing terrestrial evolution to start producing home-grown, true-blue terrestrial lifeforms.

In 1979, the famous English astrophysicist Sir Fred Hoyle and the widely respected astrochemist N. Chandra Wickramasinghe published a book (see the references) advancing an interesting variant of these ideas. They claim that the sudden appearances of widespread epidemics in modern history happen not because of the emergence of some new strain of virus or germ on Earth, but because new virus molecules or even bacteria strike the planet in comet fragments and filter down through the atmosphere, causing sudden outbreaks of disease such as Asian flu. Their theory has not been widely accepted but testifies to the continuing interest in the state of biochemical evolution of material in interplanetary debris from the early solar system.

There is strong evidence that in most meteoritic material, organic chemistry never evolved as far as any vigorous life. An amino acid molecule can have a "right-handed" or a "left-handed" structure, depending on the placement of atoms. One type is a mirror image of the other. Amino acids produced in the laboratory by nonbiological processes have a 50-50 ratio of right- to left-handedness because ordinary chemical processes have a random chance of making either type. Biology, on the other hand, keeps replicating only the type it starts with. Thus most amino acid molecules on Earth occur only in one-handedness, as if life got started from one form and reproduced that form until one overwhelmed the other.

The amino acids in meteorites have a 50-50 ratio of left-handedness to right-handedness, proving that they are not terrestrial biological contaminants that entered the meteorites after they fell, but also proving that the meteorite amino acids were synthesized in ordinary chemical processes like the Miller experiment, not replicated by lifeforms in

3500 My

2000 My

2400 My ago. Contrary to popular opinion, volcanoes were not the only landform during most of Earth's development, but they were usually the most dramatic. Active volcanism has probably always been a land-shaping agent, even if sparsely dotted over the planet.

the meteorite parent bodies. The meteorite parent bodies got only partway toward producing life.

By meteorites, I refer to rocky bodies that hit the ground. Whether meteorite data rule out virus or organisms in comets (as hypothesized by Hoyle and Wickramasinghe) is another matter. Comets are made of ice and carbon-rich dust. On the one hand, the availability of water and carbon compounds makes them better candidates for biochemical evolution than rocky meteorite material, but on the other hand, they do not leave many solid meteorites when they collide with Earth. In other words, we may not have good enough samples of comets to know what sort of organic chemistry occurred in those ancient bodies. Scientists in several countries have discussed launching a robotic space mission to get a sample from a comet, but this is probably a decade or more in the future.

At any rate, processes such as the Miller experiment certainly produced overwhelming amounts of amino acids within Earth's ecosphere. Therefore, whatever the sources of organics during the first few hundred million years, we are probably descended primarily from organic materials that formed on Earth during the tail end of the intense cratering.

ABORTED JOURNEYS IN EVOLUTION?

Even within that framework, there are interesting mysteries about the progress of evolution. First of all, are we directly descended from the first life-forms? Or did evolution produce generations of early, unfamiliar creatures subsequently wiped out by giant, ocean-vaporizing impacts or other climatic catastrophes? Are we products of Earth's first try at life, or only the second, third, or hundredth? Perhaps life could not start until conditions were relatively benign—say, 3500 My ago. In that case, we may be directly descended from such life.

But if the formation of life was as inevitable as some researchers think, then it may have started quite early—for example, 4200 My ago—only to be wiped out time and again. Each start would have led to the evolution of certain styles of primitive life-forms, filling the environmental niches that were then available. Probably none became more advanced than single-celled slimes or simple oceanic organisms, but some of these may have been very different from the forms of life that we know. According to this theory, it was only after the cratering rate dropped, the ocean stopped being vaporized in mighty impact explosions, the crust stabilized, and the climate became fairly constant, that the final chain of life began, of which we are the most recent link.

One observation supporting this second theory comes from studies of the earliest microfossils. In a 1988 conference, Michigan geochemist James Walker remarked that Earth may not have been continuously habitable even as recently as 3500 My ago. There is no proof that the more recent microfossils in the general age range of 3000 to 2800 My ago are clearly descended from the earlier fossils. Some later creatures seem quite different. This might suggest large impacts that wiped out earlier "alternative" species and made evolution start all over again.

If so, it is spooky to think about all the creatures that might have been. Would the first evolutionary experiments have eventually produced humans and other mammals, or might the later evolutionary products have been entirely different?

CREATURES OF THE MIST

A related idea involves fossils from much later in the geologic record but still early in the mists of geologic time. As we will discuss in more detail later, the abundant large, well-preserved fossils that allowed Lyell and other classic geologists around 1800 to define the stratigraphic record do not appear until amazingly recently—about 550 My ago at the beginning of the Cambrian Period. This is only an eighth of the way back in Earth's history! Apparently, it took about some 4000 My for evolution to produce creatures big enough and hard-bodied enough to leave prominent frequent fossil impressions in sediments. This fossil sequence, once established, led fairly directly to modern lifeforms.

In the last few decades, geologists have begun to recognize mysterious fossils from an earlier time. Many fossil organisms from 670 My ago until around 540 My have unusual body structure. They are animals, but they have flat structures like leaves or feathers. They may have absorbed nourishment from outer surfaces instead of ingesting it through mouths and tubular inner organs. Around 1984, German paleontologist Adolf Seilacher suggested that these were truly an alternative line of evolution that was killed off about 550 My ago by some disaster such as an asteroid impact, allowing an emergence of new animals with bulkier, tubular body forms—the pattern on which we are built.*

Some of the peculiar strains may have survived for a short time into the beginning of the Cambrian, as shown by the famous Burgess Shale fossils of British Columbia, dating from around 540 My ago. These fossils give a rich sampling of life at that time. While eight of the approximately thirty modern-day phyla of animals can be found among them, as many as ten groups are grotesquely unfamiliar. Their status is controversial. For example, one animal named *Hallucigenia* looked like something from a drug addict's nightmare: it seemed to walk on seven pairs of spike-like legs and have seven tentacles along its back. However, Chinese fossils analyzed in 1991 by a Chinese-Swedish team suggest that *Hallucigenia* was not so strange after all and may be related to a Southern Hemisphere centipede-like creature (Ramskoeld and Hou, 1991). Some of the bizarre early Cambrian creatures, however, seem to represent evolution's experiments with other body plans that failed, due either to eco-disasters or to their own inability to compete with other organisms. This supports the idea that evolution included now-lost "alternative" lifeforms.

---

*An article in *Science* (January 6, 1984) discusses these ideas. Uncharacteristically, it has a headline worthy of a supermarket tabloid: "Alien Beings Here on Earth." The point was that if we want to know how alien beings on other planets might look, we should study alternative independent lines of evolution that may have produced unfamiliar lifeforms on ancient Earth.

Hartmann

2250 My ago. Halfway through Earth's history. Eroded volcanic mountains and craters, uncloaked by vegetation, were an important landform in many continental areas, as in this aerial view. The degree of continental drift during this period, along with the consequent frequency of folded mountain ranges such as the Alps, is somewhat controversial.

CREATURES OF THE DARKNESS

There is still another type of evidence bearing on "alternative" lifeforms. During the 1970s, undersea research vessels began to find evidence of hot springs emitting geothermally heated water. In 1977, one of these was located for the first time by a piloted submarine craft. It was 2500 meters down, on a lava sea-floor region normally bleak and devoid of most life. But suddenly, as researchers John Edmond and Karen Von Damm describe it (see the references), they "came on a fabulous scene. . . . Here was an oasis. Reefs of mussels and fields of giant clams were bathed in the shimmering water, along with crabs, anemones, and large pink fish."

The extraordinary thing was not the abundance of life in the dark depths, but the ultimate source of these lifeforms' energy. The ultimate source of life-giving energy for most deep-sea creatures is the sun, because the sun allows microscopic plant forms called plankton to grow at the ocean surface; small fish of the surface waters eat the plankton; larger fish from deeper layers eat the small fish; and so on. On the basis of such observations, biologists have long stated that all life, except for certain mineral-consuming bacteria, ultimately depends on the sunlight. On the dark sea floor, far below the fertile, sunlit layers, creatures were known to be few and far between, and even *they* were thought to depend ultimately on the sun.

Why, then, was there so much life crowding around a dark sea-floor vent? The answer was that hydrogen sulfide (H_2S), a common volcanic gas, was being emitted along with other gases and fluids from the vents. Bacteria had evolved here that derived energy from chemically breaking down the volcanic hydrogen sulfide. Instead of depending on photosynthesis, they use chemosynthesis to create nutrients. The abundant bacteria were an attractive food source for other sea creatures, which explains why colonies of more familiar creatures had gathered in the darkness around the vents. The unique aspect was that the whole community depended ultimately not on the sun but on geothermal energy.

Many similar vents were eventually found, spewing out mineral-rich geothermally heated water at temperatures reaching more than 700°F. The minerals were dissolved in the hot water on its way upward through Earth's undersea crust, but condensed when they hit the cold ocean-floor water, creating deposits of iron, zinc, and copper sulfides and also calcium and magnesium sulfates. Among mineral deposits, a strange biological niche existed around each deep-sea vent. The species that gathered there included not only relatively familiar animals, such as giant clams (a foot across), but also unfamiliar species that had evolved to fit the peculiar conditions of the night-dark sea-floor oases. The latter include twelve-foot-high vertically oriented tube worms that exist in symbiosis with the chemosynthetic bacteria, which live inside the worms' bodies.

The important thing is that for the first time biologists can study colonies of lifeforms that do not depend ultimately on sunlight. In the blackness of the ocean depths, surrounding the smoking volcanic vents, the bacteria that depend on hydrogen sulfide instead of sunlight are truly alternative terrestrial lifeforms. Almost as isolated as aliens on other planets, they have followed their own sun-independent evolution. They increase our estimates of the chances that life could have started in a wide variety of environments, either on primitive Earth or on other planets and moons normally dismissed as too far from the sun to sustain life. If such species exist in isolation on today's Earth, who knows what weird species involved in unknown microenvironments of the past 3000 My?

THE PERIOD OF RNA LIFE—3500 TO 2800 MY AGO

Still another line of evidence suggests that life on Earth has tried unfamiliar experiments. In the main line of biological evolution, we encounter unexpected chemistry along the way. In modern cells, certain molecules called enzymes promote the

3500 My 2000 My

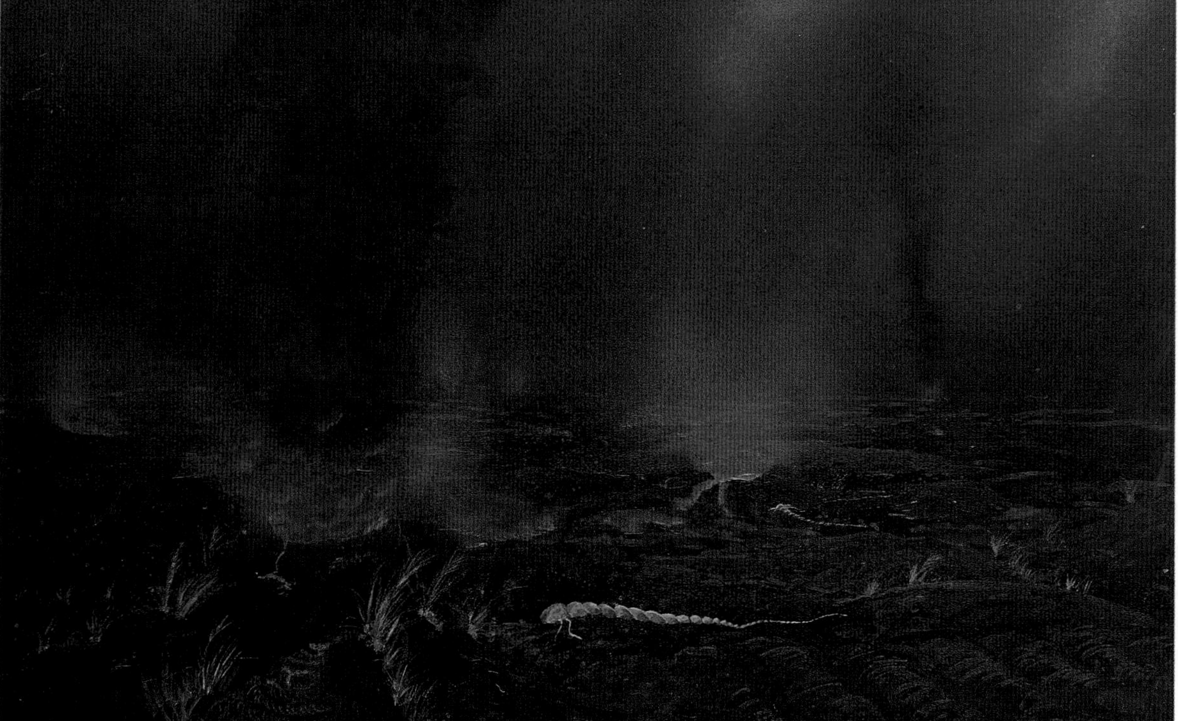

2000(?) My ago. Some scientists believe that unfamiliar organisms may have evolved in the ancient oceans, only to be completely wiped out by large impacts that restarted the chain of evolution. This imaginary reconstruction shows some "alien" Precambrian species clustered around volcanic vents on the ancient sea floor.

chemical reactions that allow the cell to thrive. In ancient cells of about 2500 My ago, these enzymes were apparently absent, and the chemical reactions were "managed" instead by the RNA nucleic acid molecules. Some biologists have called this ancient era the "RNA world" because its basic cellular operating chemistry was different from that of our world.

The whole idea of an RNA world on early Earth has been controversial. It was discussed as early as the 1960s by the researchers who unraveled the molecular structure of RNA, DNA, and the gene. The theory of a primeval RNA-dominated world helped explain a chicken-and-egg paradox about DNA and proteins in the cell. Protein molecules play a role in the repair and reproduction of DNA molecules, but DNA contains the genetic information that helps cells manufacture the proteins in the first place. Which came first? DNA controls most cells now, but how could the DNA itself have got started? A preliminary stage in which RNA carried out the cell functions and helped create both DNA and proteins at the same time would explain the problem. So the idea of an unfamiliar, primordial RNA world, around 3000 My ago, became popular among biologists by the early eighties.

Some theorists speculated that the RNA world did not consist of only a few poor, simple lifeforms but instead was full of vigorous organisms. A Swiss-American team of molecular biologists in 1989 pictured a rich world of diverse RNA lifeforms, eventually evolving to produce one "breakthrough organism" that began to facilitate its cellular reactions using the modern enzymes in conjunction with DNA, as well as RNA, molecules. Thus, in this view, DNA, proteins, and we ourselves are descendants of strange RNA lifeforms that eventually died out because they couldn't compete with the lifeforms that evolved after the "DNA revolution."

Other scientists in 1989 questioned the existence, or at least the vigor, of the RNA world. Experiments showed that RNA is hard to synthesize in a test tube or in early nature because it is so complicated. So more and more scientists asked how RNA itself formed in the first place. At the very least, the defenders of the RNA world had to admit that some pre-RNA step has gone undiscovered.

In summary, an area of current excitement at the interface of evolution, biology, and molecular chemistry is the question of the kinds of organic molecules and living creatures that existed at the beginning of biological evolution. All the ideas discussed above support the concept that primordial life was unfamiliar. Perhaps the next generation of scientists will shed more light on the earliest organic chemistry of Earth and the strange biology it produced 3000 My ago.

AN ALIEN DAY, AN ALIEN NIGHT

Recall from Chapters 5 and 6 that the moon has been moving outward from Earth ever since it formed, and that the day has been lengthening as a result.

Interestingly enough, Earth's ancient shorter day can be confirmed from some of the earliest fossils and geological deposits from as long as 2800 My ago. For example, changes in particle size in sediments or growth rings in organisms may record such cycles as annual seasonal variations in floods, daily tides, and so on. From such data, scientists can reconstruct ancient data such as the number of days in a year.

By 2800 My, the date of some of the earliest records of tidal and monthly cycles, the moon had moved most of the way to its present position. Still, the day's length and the size of the moon in Earth's sky were noticeably different at that time. From estimates of the number of days (Earth rotations) in a year (the time for Earth to go around the sun), we can estimate that 2800 My ago there were almost 500 Earth rotations in a year. This means the day was about three-fourths as long then as now, or about eighteen hours long. This in turn corresponds with the moon being at about 86 percent of its present distance, meaning that the moon looked roughly 16 percent bigger in diameter in the sky than it does today. Solar eclipses, while still rare, were correspondingly more frequent and lasted longer than they do today, and the moonlit night sky was brighter.

The night sky was strange in another way. The cluster of stars that formed along with the sun had long since drifted apart, so that the night sky no longer had an unusual number of bright nearby stars. Nonetheless, the sun and its attendant solar system of planets takes about 230 My to go around the galaxy, and this means that 2800 My ago the sun had completely different neighbor stars than it does today and was probably in a different part of the galaxy altogether. In short, the night sky had about as many stars as it does today, but in totally different patterns of constellations.

Thus even the seemingly familiar moon and stars conspired to make a day and a night on Earth weirdly, subtly different in the era when strange lifeforms were taking their first tentacle-like grip on the planet.

3000 to 1000 My Ago

9. Life's Slow Progress and a Radical Change in Earth's Atmosphere

> I seemed to sip the silver flavors of a time as ancient as the starlight sprawling on the grass.
>
> —*Charlie Smith,* Shine Hawk, *1988*

As we have seen, life probably started as a result of the tendency of proteins and other giant organic molecules to aggregate into microscopic spherules. The transition from inert spherules to true cells must have involved nucleic acids, such as DNA, with their ability to replicate themselves. The structure of the DNA is the famous double helix, similar to necklace strands twisted together. DNA can reproduce itself when the two

strands come apart. That is, more amino acids can attach themselves along the line of the break, re-creating the original double-strand pattern.

DNA has another special quality. It is so huge that even segments of its long structure are very complex—complex enough to control chemically the replication of various traits not only of the "daughter" molecules but of the whole cell. The way in which this works is complicated and not fully understood.

To get a glimmer of it, we have to go back to the molecular nature of things and give an example at the simplest level, among the protein molecules. Amino acids string together to make long, chain-like protein molecules. Most protein molecules contain only twenty different kinds of amino acids, but the acids may repeat in different groups, to make enormous proteins. In general, the individual amino acids have hydrogen atoms and other small groups of atoms sticking out on different sides, and these can form bonds in different directions. The *sequence* of amino acids therefore determines the

3000 My ago. A barren landscape marked much of Earth's history. Volcanic hot springs and bubbling mud pots pour steam (water vapor) and carbon dioxide into the early atmosphere in this view.

shape that the protein can assume, and how it in turn can interact with other materials. For example, in one amino acid sequence, hydrogen bonds form between every fourth amino acid and hold the chain in the form of a helical coil. In other arrangements, the hydrogen atoms form bonds between different chains, holding them side by side in sheet-like arrangements.

These examples at the molecular level hint at how the complexities of organic molecules can control the shape of larger structures in which they form a part. Remember that the proteins in turn are building materials for cells and tissues. Proteins can form hemoglobin, collagen (the chief constituent of connective tissue and of bones), and other crucial materials. Proteins and DNA work together to control the familiar structures of plants and animals, regulating everything from the texture of your hair to the shape of your nose.

In modern, advanced lifeforms, DNA in each cell is concentrated in the nucleus—hence the name nucleic acid (for which the N and the A stand). In advanced organisms, the DNA is grouped into bodies called chromosomes in the nucleus of the cell. Certain segments—that is to say, certain sequences of amino acids along the twisted chain of the DNA molecule—control reproduced traits. These segments are called genes. Genes are the molecular-level keys to how the cells will be built. A particular gene may lead to construction of hair that is curly, but a different set of amino acids in that part of the DNA molecule might produce hair that is straight.

DNA transmits the "instructions" about how to build the protein structure of an organism via the genes. A DNA molecule in a cell nucleus breaks apart. Each half forms a long chain of amino acids that might be likened to one side of a zipper. In this analogy, each metal cleat represents an amino acid group, and a sequence of metal cleats makes a particular gene. Each metal cleat in this region can attract other molecules, forming the other side of the zipper, which is also a long chain molecule. Instead

Miller

3000 My ago. Life gains a foothold on rocks in tidal pools. Life had not yet colonized the land, but lichen-like slimes and molds might have begun to get a toehold on rocks around edges of tidal pools. Volcanic activity, as shown in the background, might have added heat and selected minerals to aid in creating the earliest biological niches.

3000 My

1000 My

Miller

2500 My ago. Stromatolites, cabbage-like colonies of algae, visible in intertidal zones, were one of the main evidences for life in Earth's landscapes for most of our planet's history.

of attracting the same amino acid groups that constitute it, thus replicating itself, a DNA molecule can form a template molecule, as discussed in Chapter 7. Certain cleats can bond with molecules not in the DNA sequence but in the RNA molecule sequence. This yields an RNA molecule as the other side of the zipper. Thus the sequence in the manufacture of proteins is: A segment of DNA makes an RNA molecule. The RNA now breaks away and becomes a template—one side of a zipper. Each of its segments now attracts a specific amino acid, so that a long chain of amino acids forms as the other side of the zipper. That chain breaks away from the RNA and is a protein molecule, with its own specific shape properties. The protein may become part of the cell wall or of some other structure of the cell, or it may serve as an enzyme, a molecule that itself helps manufacture still more organic substances.

In the language of molecular biologists, the DNA molecule stored the instructions to build a template for the manufacture of specific materials used by the organism. The DNA bundles are called chromosomes, and specific sequences of amino acids along the DNA—the sequences that control the structure of certain parts of the organism—are called genes.

The important point is that the original patterns, or instructions, are stored in the DNA of the chromosomes in the nucleus of each cell.

Of course, modern, advanced lifeforms are so complex as to seem almost magical to us. It is barely imaginable how these molecular processes, cumulatively, could yield such complex masses of organic material as, for example, a chimpanzee, with its unique behavior. But when thinking about the history of life on Earth, we need not visualize only the finished product from our macroscopic point of view: complex behavior patterns such as the mating of bears or the feeding of a Venus's flytrap. Instead, we need to think about the simplest, molecular basis for life, which drives everything else.

British biologist Richard Dawkins goes so far as to suggest that we are mistaken in thinking of ourselves and other animals as the basic entities of life, the masters of the world. Rather, it is the genes, or reproducing molecular units, who are running things. We are just the giant, lumbering machines that genes, in 3500 My of biochemical evolution, have constructed to carry them around and allow them to interact with other genes and reproduce themselves. (Dawkins's book *The Selfish Gene* is listed in the references.) Different chemistries in the genes lead to different design plans for the lumbering machines, and the machines that are well adapted to environmental niches reproduce the genes most successfully. Those species live on. Other machines tend to die out. In this sense, life is really the process of complex molecules aggregating and then finding other complex molecules with which to join in order to keep reproducing themselves. In 3500 My, the lumbering machines have grown very complex!

WHAT CELLS TELL US

Let's try to use the simplest organisms of today to figure out what was going on 3500 My ago, when the first known substantial organisms—single-cell lifeforms—were beginning to appear. Although single-cell lifeforms are still the most abundant lifeforms on Earth, they weren't even discovered until the Dutch naturalist Antonie van Leeuwenhoek (1632–1723) used early microscopes and identified them in 1677; only after 1677 could lifeforms be studied and classified according to their microscopic cellular structure.

Such study revealed an interesting fact. Among some bacteria and algae are very primitive kinds of cells similar to fossil single-cell lifeforms of 3000 My ago. Their structure is much cruder than that of our cells. They have no nuclei. Simple nucleic acid structures exist, but they are not walled off from the rest of the cell by the kind of membrane that surrounds the nuclei in our cells. These cells are called *prokaryotes* (from Greek roots meaning "before the nucleus").

From prokaryotes we get a tremendous clue about the nature of early single-cell lifeforms. Although a nucleus is absent, a clump of DNA may be more or less concentrated in the center, testifying to the importance of nucleic acid molecules in controlling cell structure. The cell is surrounded by a tough, complex membrane that may be secreted from the cell itself. The membrane in some cases carries out photosynthesis—the process of deriving energy for the cell from sunlight.

Two major types of prokaryotes still exist in the world around us: the microscopic *bacteria* and their close relatives the *blue-green algae*, which aggregate into seaweed-like clumps.

The theory that prokaryote cells were the first lifeforms is based on two strong bits of evidence. First, they are so much more primitive than other lifeforms in the world today. Second, the most ancient microfossils resemble bacteria and blue-green algae.

The fact that blue-green algae cells grow in colonies turns out to be extremely important in understanding life before 2000 My ago. The colonies are matted, pillow-like structures one or two feet across, often spread across shallow sea floors like different-sized cushions thrown off a couch. These lumpy algae colonies are called *stromatolites*. Their importance is that they form recognizable fossils. As early as 1954, the geologist S. A. Taylor and the paleobotanist E. S. Barghoorn showed that the Gunflint Chert, a deposit of flint-like rock in Minnesota and Ontario, was made up of microscopic fossils of algae and bacteria-like forms. They were stromatolite deposits, about 1900 My old. By 1977, fossil stromatolites older than 3000 My had been found!

The proliferation of these ocean-dwelling algae colonies around the globe by 3000 My ago confirms the vigor of early life in the oceans.

Remember once again that the classical geologists were able to establish a fossil sequence back to only about 550 My ago. That was because hard-bodied lifeforms, those capable of leaving fossils, emerged only that recently. Therefore, until the 1980s, geology books discussed the biological and geological evolution primarily of only the last 550 My—the last 12 percent of Earth's history.

The stromatolite discoveries, however, show that abundant algae colonies prospered in the shallow seas of our planet long before the more familiar hard-bodied lifeforms. Not only were they among the first successful lifeforms, but they have also thrived successfully until the present, where despite some problems with today's oxygen-rich atmosphere (see below), they prosper oblivious to the "improvements" and complications that have gone on around them! They can still be seen in various locations, such as parts of the Australian coast.

1500 My ago. The rainier parts of Earth must have been carved into gullied landforms during most of Earth's history because there was no plant cover to protect the soil from erosion. Rainfall runoff sculpted mountains and transported their soils downhill, depositing them in broad, gravel-filled plains.

A MORE ADVANCED TYPE OF CELL—2500(?) MY AGO

Another type of cell eventually appeared, the ancestor of our own cells. Two major advances marked its departure from the old prokaryote cells of the algae and bacteria. First, a membrane evolved to separate the nucleus from the rest of the cell. The nucleus became a sort of master cell inside the cell. Second, the nucleic acids such as RNA and DNA were more tightly organized into chromosomes, small clumps of nucleic acid molecules inside the nucleus. These types of cells are called *eukaryotes* (from Greek roots denoting a well-defined nucleus).

No one fully understands the steps by which the eukaryote cells evolved, or the reasons they were so much more successful than the algae and bacterial prokaryotes, with their vague nuclear structure. One theory is that they came from internal evolution of the prokaryotes alone. Another theory is that of Lynn Margulis, the biologist who stresses the symbiotic nature of structures within cells. She believes that eukaryote nuclei evolved when other small molecular bodies or organisms invaded prokaryotes and then evolved to the role of nucleus. In any case, once eukaryotes appeared, they were much more efficient at responding to the evolutionary impetus of natural selection. We infer this because they eventually exploded into a variety of organisms, not only single-celled but also many-celled. Harvard biologist Ernst Mayr calls their first appearance "the single most important event in the history of the organic world" (Mann, 1991).

We don't know exactly when they first appeared. From microfossils alone, it is hard to distinguish the earliest eukaryotes from prokaryote cells. Perhaps they emerged between 3000 and 2000 My ago.

There were two broad groups. One resembled altered forms of algae and evolved into plants. The other, more complex group, called protozoa, evolved into animals.

Hartmann photo

Earth's atmosphere is a dynamic system, including turbulent clouds at lower levels, and high-lying, more sheet-like clouds, as seen in this view from the summit of 14,000-foot Mauna Kea volcano, Hawaii.

We can sense that the long march that led to the eukaryote was not an easy one: Earth used up half its history inventing the eukaryote cell. It took an equal amount of time to get from that cell to us. In fact, the most explosive change in eukaryote evolution came amazingly recently, in terms of the planet's history. This was a proliferation of many different species of larger, *multicelled* organisms, both animals and plants, only about 700 My ago. The reason such a proliferation came so late is uncertain. Some scientists suggest that it was a consequence of a fantastic change in the environment of Earth.

THE MOST DRAMATIC CHANGE IN THE HISTORY OF EARTH'S ENVIRONMENT—2200 MY AGO

As emphasized earlier, the early atmosphere of Earth was mostly carbon dioxide and water vapor. This situation lasted until about 2300 to 2000 My ago. Then the evolution of life changed everything. Remember that the earliest lifeforms 3500 My ago included blue-green algae; other single-cell algae forms and other ancestral plants appeared as soon as prokaryotes evolved. These organisms all conducted photosynthesis. This means that in order to build their organic molecules with the aid of solar energy, they acquire carbon by breaking down carbon dioxide molecules from the atmosphere. Thus carbon dioxide is consumed by the algae or plant cells, the carbon atoms are incorporated into cellular molecules, and the oxygen atoms are emitted into the air.

The astonishing thing is that this modest process, conducted by tiny creatures for a thousand My, changed the entire Earth. Gradually the carbon dioxide was depleted and the oxygen accumulated until it became a major constituent of air.

WHAT THE AIR TELLS US

We think of Earth as an oxygen planet. That is somewhat misleading, because the present atmosphere is 80 percent nitrogen and only 20 percent oxygen. Nonetheless, oxygen planet is an apt name because oxygen is one of Earth's most extraordinary features.

The oxygen content of our atmosphere is extraordinary because oxygen is one of the most chemically reactive gases. For example, it is oxygen that combines with iron to cause rust on metal, reacts with the chemicals in a match head to make flame burst forth, and combines with gasoline to run a motor or cause an explosion. Oxygen is so reactive that it would not last long in a planet's atmosphere unless it was constantly replenished. It would simply combine with hydrogen to make water, or if there was inadequate hydrogen available, it would combine with blackish iron-rich minerals in rocks—minerals whose iron is in the so-called ferrous state—to form other reddish, rusty, rock-forming minerals whose iron is oxidized, or in the so-called ferric state.

On Earth today, a similar process creates geological strata called red beds. The reddish sedimentary rock layers of the Grand Canyon and other landscapes of the American Southwest are a famous example. Red-bed formation is very effective in fairly arid regions, with alternating moist and dry cycles, such as in the Southwest.

In summary, if you are cruising around the universe in your hyperdrive spaceship and you find a planet with oxygen in its atmosphere, you can be pretty sure that something strange is happening to keep the oxygen replenished. That "something strange" will probably be life.

ARCHAEOLOGISTS OF THE ATMOSPHERE

Photosynthesis has been vigorous for only about half of Earth's history, and advanced plants have flourished for only the last eighth of its history.

For these reasons, we can be sure that in the first half of Earth's history, before life was well evolved, there was much less oxygen than the 21 percent (as measured by volume—the way the figure is usually given) that characterizes today's air. Scientists began to suspect this as early as the mid-1800s. Modern researchers estimate that the air of primordial Earth had less than 1 percent of the present oxygen supply, and not until about 2000 My ago did the oxygen supply reach even 1 percent of its current value! (These numbers are based on studies in the 1980s by American, Soviet, and other researchers. See Budyko, Ronov, and Yanshin, 1985, p. 23ff.)

Thanks to expanding research on the evolution of air, we can be more specific about Earth's atmospheric history. Let us recap what we've said so far.

Earth probably started its existence with a first-stage atmosphere that was merely a concentration of the hydrogen and helium of the surrounding interplanetary nebular gases. This was probably blown away almost as soon as Earth formed, because of giant impacts and the solar wind gases that rush out from the sun. The second-stage atmosphere, as we have also discussed, was the extremely dense carbon dioxide (CO_2) and water vapor that fumed out of the early volcanoes. The water vapor condensed to form the oceans, and for the first thousand My or so, Earth had a dense CO_2 atmosphere.

The great density and pressure of the second-stage CO_2 atmosphere declined even before plants got busy, because CO_2 dissolved in ocean water, making a weak solution of carbonic acid.* The weak carbonic acid reacted with rock minerals of the sea floor to make carbonate minerals and rocks, using up the CO_2 and trapping it in the newly formed carbonate rocks. As the first CO_2 in the ocean water was converted to carbonate rocks, more CO_2 was absorbed into the ocean from the atmosphere.

In other words, CO_2 from the air was cycled through the oceans and stored in carbonate rocks. Although the original second-stage atmosphere of perhaps 4400 My ago had about seventy times as much pressure and density as today's atmosphere, the pressure by 2500 My ago had declined toward today's values. It was in that environment that plants began to consume the remaining CO_2 and replace it with O_2.

EXPLAINING THE FIRST ANCIENT ICE AGE—2300 MY AGO

Some of Earth's earliest known widespread glaciers date from 2300 My ago. They are similar to the ice sheets that covered much of North America and Europe during the "recent" ice ages, but they lasted much longer. Glacial deposits dating from 2300 My ago are found on rock units of many continents. Why did ice spread around Earth then but not before?

*CO_2 dissolves easily in water. Such a solution is what makes your soft drinks fizzy.

A rough sketch of Earth's continental landmasses 1100 My ago. Most of the continental masses are believed to have been near the equator or in the Southern Hemisphere. This and following sketch maps are based on a variety of sources; the uncertainty is high for the earliest dates. North and south poles of Earth's rotation are at top and bottom, respectively, where ice caps would form.

The answer may lie in the emergence of the early, simple plants and their consumption of CO_2. Because CO_2 was more abundant in the first half of Earth's history, a CO_2 greenhouse effect prior to 2300 My ago probably kept the temperatures in many areas above freezing, even though the sun was less luminous then.

Paleontologists speculate that once simple algae and other plant life began prospering and consuming CO_2, a drop in the CO_2 content of the air occurred. This may have triggered a drop in temperature over a period of several hundred My. Fortunately, this happened at a point in history when the sun was about to become luminous enough to keep most latitudes above freezing. Thus, while the glaciers came on fairly rapidly (geologically speaking) with the drop of the greenhouse effect below a critical value, they gradually retreated again as the sun continued to warm up, bringing Earth to a milder, unglaciated climate again only a few hundred My later.

DIGRESSION: WHAT VENUS TELLS US

Earth's sister planet, Venus, gives us very important clues about this story. Venus is our next-door neighbor, nearly the same size as Earth and the next planet closer to the sun. The surface of Venus has many volcanoes and lava flows. But it lacks oceans, and instead of a nitrogen-oxygen atmosphere it has an extremely dense atmosphere of carbon dioxide. Why? Its volcanoes probably once produced a water-vapor atmosphere like early Earth's, but the water vapor could not condense because of the greater heat from the nearby sun. With no condensation, there was no rain on Venus, and no oceans. The water vapor molecules eventually dissociated due to sunlight and escaped from Venus. With no oceans, Venus's carbon dioxide could not form carbonic acid, attack rocks, and get stored away in the form of carbonates. In other words, Venus still has a perfectly preserved second-stage atmosphere—an actual example of the dense CO_2 atmosphere hypothesized for Earth. This strongly supports our theories of Earth's atmospheric history.

Even the total amount of Venus's atmosphere checks with the theory. Since the two planets are about the same size, we would guess that they both produced about the same amount of CO_2 from their volcanoes. The total amount of CO_2 on Venus has been measured at about seventy times the total amount of air on Earth. If you took the total amount of CO_2 "trapped" in the carbonate rocks of Earth and put it into the atmosphere, you would get this same amount of gas, an atmosphere about seventy times as dense as today's. You would have an atmosphere very much like Venus's and would have re-created Earth's second-stage air.

2300 My ago. An early freeze. One of the earliest known periods of widespread glaciation may have been associated with emerging plants' consumption of carbon dioxide at that time.

Hartmann photo

Profile of a storm cloud. Updrafts of warm air carry moisture to high levels, where it condenses as clouds and raindrops. Rain falls out of the lower part of the cloud while a giant anvil-shaped updraft forms above. Grand Canyon, Arizona.

EVIDENCE FOR THE TRANSITION TO OXYGEN

Emergence of Earth's familiar oxygen atmosphere only in the last half of Earth's history may seem like a wild idea based on grandiose theories. Not so. If the evidence from Venus does not convince the skeptic, there are several direct terrestrial observations that indicate that there was little oxygen before 2200 to 2000 My ago, and that the oxygen content rose past some critical level at that time.

The air is a very tenuous medium, so you might suppose that it would be hard to learn anything definite about its early properties. But the archaeologists of the atmosphere have made direct observations on Earth that confirm the absence of oxygen in the early air. These are geological observations that made no sense until they were linked with the idea of a rise in oxygen content.

One line of evidence was first pointed out by the Canadian researcher S. M. Roscoe in 1969 (see Cloud, 1988). He noted that in deposits now dated at 2450 My, certain minerals that could have easily been oxidized were deposited without oxidation. Furthermore, red-bed strata—the ones produced by oxidized mineral deposits—are rare before 2200 My ago. Rock sediments from before 2000 My ago are commonly black or gray. A similar story is told by sulfur isotope chemistry. The evidence points to the beginnings of oxygen accumulation in the atmosphere around 2200 My ago.

The geological evidence implies that considerable time was required for the oxygen level in the air to increase. Oxygen is so reactive that the first oxygen released by plants was quickly consumed in making oxidized minerals. This may be why the evidence for the most rapid buildup of oxygen comes not at 3500 or 3000 My ago, when plants began to prosper, but later, around 2000 My ago.

In fact, if we had been able to land a space probe any time prior to 2000 My ago, we would not have recognized an atmospheric sample as coming from Earth. The carbon-dioxide-rich sample would have looked more like the composition of the atmosphere of Mars or Venus. The present atmosphere, of roughly four-fifths nitrogen and one-fourth oxygen, by volume, can thus be characterized as a third-stage atmosphere in the history of our planet, and the emergence of oxygen as a major constituent as the most radical change in the history of the atmosphere's composition.

LIFE AFFECTS ATMOSPHERE AFFECTS LIFE

Still another line of evidence for the late emergence of oxygen is the proliferation of oxygen-consuming species of life only late in Earth's history.

This involves a striking aspect of the great change from CO_2 to O_2: the feedback effect associated with the emergence of oxygen-loving lifeforms. Life caused the atmosphere to change, but then the atmospheric change caused life itself to change. *The earliest organisms could arise and evolve only in an oxygen-poor environment.* Because oxygen is so reactive, it is really somewhat dangerous for life. Without adequate protection, early organic molecules and organisms would have been damaged if exposed to the oxygen levels we encounter. Even today, stromatolite-creating colonies of blue-green algae thrive best in oxygen-poor environments.

But after the oxygen levels began to increase, newly evolved organisms could make use of the very reactive gas to enhance their energy-producing activities. In fact, as we know, lifeforms proliferated madly and evolved to much more complex states by utilizing the very same oxygen-rich environment that would have wiped out their own ancestors. Some paleontologists believe that the eventual proliferation of animal life 700 My ago was possible only *because* the plants had changed the atmosphere. It is remarkable to realize that this mad proliferation was concentrated primarily in only the last eighth of Earth's history.

BOOK FIVE

Earth Matures

1000 to 400 My Ago

10. The Flowering of Life

'Tis very probable that divers . . . Transpositions and Metamorphoses have been wrought even here in England. . . . There may have been divers Species . . . wholly destroyed and . . . others changed. . . . There may have been divers new varieties generated.

—*Robert Hooke (1635–1703), presaging Darwin, in* The Posthumous Works of Dr. Robert Hooke, *1705*

Even though the atmosphere and climate of Earth were more or less familiar by 1000 My ago, with nearly the modern oxygen content, lower CO_2, and a declining CO_2 greenhouse effect, the land itself was still unrecognizable. In the first place, continents had not yet assumed either their modern shapes or their modern positions. No modern coastlines or familiar mountain ranges yet existed.

At times, ice sheets tied up large amounts of water, blanketing continents and lowering sea levels; we are coming out of such a period now. During much of the last 1000 My, the climate was somewhat warmer, ice fields were smaller, and sea levels were higher, so that continents were often flooded by shallow seas. Furthermore, the

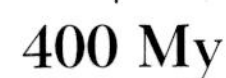

About 500 My ago, the continental masses formed two major areas: Gondwanaland (in the Southern Hemisphere) and Laurasia (Northern Hemisphere). Separating them is the Tethyan Sea. Parts of Africa experienced Antarctic conditions.

impermanent continents are parts of large crustal blocks, which drift across Earth's surface, collide with each other, even sink into the mantle. They can change shape dramatically within only 50 million years. For all these reasons, familiar landmasses such as North America were not recognizable 1000 My ago.

ANCIENT GLACIERS—850 MY AGO

According to geological studies, many regions of the present continents were under ice-age glacial conditions in the interval from about 950 to 650 My ago. These are not the Ice Ages we learn about from schoolbooks; the Ice Ages happened only tens of thousands of years ago when the first modern humans were painting mammoths on cave walls. These are much earlier ice ages, long before humans existed, and evidenced by signs such as deposits of ancient glacier-transported gravels, dropped by glaciers as they melted. Such signs show that periods of glaciation were unusually frequent on many continental surfaces from about 900 to 650 My ago. This period has been called "the longest winter."

Recalling that the continents themselves drift around, can we explain the ancient ice ages by assuming that glaciers formed on continents only when the ancient continents were near the North or South Pole? Does that explain the long winter of these continents? Apparently not, if we believe certain magnetic studies of the ancient rocks. These studies place the ancient landmasses fairly near the equator at the time the glaciers existed! If the magnetic data are right, "the longest winter," of 950 to 650 My ago, probably involved a global cool spell in which even low-latitude continents had extensive glaciation in their mountains.

WHAT MAGNETISM TELLS US

How do magnetic studies locate ancient continents? Such studies are worth talking about in their own right. They detect the orientation of Earth's magnetic field around the rock as it formed. Recall that a compass needle aligns itself with Earth's magnetic field. Generally, at low latitudes this is nearly horizontal, parallel to the ground. But if the needle were free to swing in three-dimensional space, we would see it tip toward the ground at high latitudes. In northern Canada, near the magnetic pole, it would point downward at a steep angle. A property of rocks allows geo-sleuths to take advantage of this situation. Within molten lava, small magnetic crystals, so-called magnetic domains, are free to orient themselves along the magnetic field. As the molten lava solidifies into rock, the domains are frozen into position. Hence, if the rock has remained fixed in its bedrock position, the orientation of the magnetic domains in the rock—horizontal or dipping at a steep angle—hints at its latitude when it formed, relative to the magnetic pole of Earth.

Because the measurements are tricky, some geologists have questioned the reliability of the

800 My ago. Precambrian hot springs on early Earth were deeply colored by algae and other microorganisms. Life had not yet advanced onto dry land surfaces.

magnetic data for deriving precise continental positions. They argue that a global cool period would be a puzzle, and that continental glaciation more likely happened only as a continent drifted near the North or South Pole. This is an unresolved controversy that may be clarified as magnetic techniques improve by the turn of this century.

EMPTY LANDSCAPES

In addition to widespread glaciers, another peculiarity of Earth 850 My ago is dramatic but easily forgotten. Whatever landmasses were present, they were almost completely barren! No herds of roaming animals, no leafy forests, no flowering fields, no tropical palms graced those ancient shores. The mists of morning did not move through trees, nor were they broken by the songs of birds or the chirps of insects.

Why is it so astonishing to realize that for most of Earth's history, an alien visitor to our planet would have found landscapes as barren as those of Mars? Perhaps it is hard for us to give up our 3000-year-old concept that Earth has flourished with plants since the third day of Creation, with animals and humans appearing on the sixth day. Human history supposedly filled up all the time since. We get a certain queasy feeling when we learn from fossils and isotopes that the land was empty for 80 percent of Earth's history; not to mention that land animals appeared only in the last 10 percent and mammals only in the last 1 percent. Is it our puritan heritage, our Western work ethic, that whispers to us uneasily about all that time being "wasted"? So much time!

But nature and chemistry required 4000 My to do the work that God is rumored to have done in the first six days. Life on the land is an astonishingly recent phenomenon!

PLANTS VS. ANIMALS

When we walk around outdoors today, we find the landscape bursting with plants and animals, and we have no problem placing lifeforms in one of those two categories, whose definitions seem straightforward. Plants are commonly fixed in place, get their energy from sunlight, and get much of their material substance by breaking down inorganic raw chemicals, such as CO_2 from the air or nitrogen compounds from the soil. Animals are generally mobile, get their energy from chemical reactions while processing food, and get their material not from raw chemicals but by ingesting other biological materials, such as plants or other animals.

Before 700 My ago, life remained fairly primitive and these distinctions were not so clear. For example, the bacteria, which are prokaryotes (lacking a well-defined nucleus), show a variety of forms. Some take in chemicals and depend on atmospheric CO_2, like plants. Others take in organic material for food, like animals. Sulfur-oxidizing bacteria, as we saw in Chapter 8, don't need light; they can prosper on powdered sulfur, ammonium sulfate, water, carbon dioxide, and other chemicals. While bacteria were traditionally classified in the plant kingdom, more recently they have been classified as a separate, prokaryote, kingdom.

By 700 My ago, however, distinctive plant and animal forms were beginning to develop in the seas. Only a few were substantial enough to leave good fossils, but they were crucial to our understanding of the planet's history. The earlier cabbage-like stromatolite algae colonies and various microorganisms left fossil imprints as far back as 3000 My ago, but they are hard to recognize, and not distinctive enough from one era to another to serve as good age indicators. What geologists needed in order to map the march of time were distinctive plant and animal species, preferably hard-bodied, and changing from one geologic period to the next. Such organisms finally appeared in abundance in the amazingly short 150-My period from 700 to 550 My ago. It was one of the most concentrated biological revolutions of all time.

THE EARLIEST "PROMINENT" FOSSILS—700 MY AGO

During this interval, at last, substantial creatures floated in the seas and crawled the sea floors. They left clear fossil imprints big enough to be seen clearly with the unaided eye. This development was noteworthy not only to fossil-seeking scientists. It is a milestone for every Earth dweller. It marks the period when Earth finally began to seem inhabited.

The fossil-forming process that preserved substantial creatures is easy enough to understand. The plant or animal body settled into fine sediment and was usually covered by other fine sea-floor sediments, perhaps deposited as rivers emptied mud into the sea. Millions of years later, these layers found themselves thousands of feet below the surface, being compacted by tremendous pressure into rocks. Even if the soft parts of the plant or animal remains had rotted away, harder parts or a simple impression could have been preserved in the now-solid rock. The rocks could have been further solidified by gentle heating from volcanic sources at still greater depth. In some cases, hot water invaded the soil layer, bringing dissolved minerals. The hot water dissolved hard parts left by the plant or animal, but

530 My ago. Some of the strangest Cambrian sea-floor organisms were well preserved in the Burgess Shale of Canada when underwater mudslides engulfed them and recorded imprints of their soft tissue. A member of the odd species known as *Hallucigenia* is shown being buried by mud.

then evaporated and left mineral deposits. Thus, instead of being simple imprints, some fossils may be crystalline reconstructions of the plant or animal. Chemical reactions could even alter some of the original body parts, leaving different minerals in the place of bones, fleshy portions, etc.

Much later, after creatures invaded the land, other types of fossils could form. For example, insects got stuck on blobs of sap, as on flypaper, and were encased inside as more sap seeped onto them. Later the sap hardened into the substance called amber, perfectly preserving prehistoric insects more than a hundred million years old. Such examples were described as early as 1729 in England. Similarly, some creatures were partially preserved when they fell into tar pits, where oily compounds bubble to the surface. A famous example is the La Brea site near downtown Los Angeles. In much more recent times, as noted in the opening chapters, mammoths and other creatures were frozen and preserved in arctic tundra.

As described in Chapter 1, fossils were the clues that allowed geologists around 1800 to begin to identify the sequence of Earth's rocks. Those geologists could map the past only as far back as the oldest fossils. For this reason, it is important to understand the important interval from 700 to 550 My ago, when the first complex fossils began to appear.

For nearly two centuries, geologists believed that the onset of the appearance of fossil shellfish and other prominent fossils at the beginning of the Cambrian Period, 550 My ago, marked the sharp beginning of the continuous chain of advanced fossil lifeforms. Everything before 550 My ago was called the Precambrian. To an earlier generation of geologists, the Precambrian seemed a vast, dark abyss of time, without known fossil markers.

The entire story of this book so far has been about the Precambrian. Geologists of only one or two generations ago hardly dreamed of filling in so much information about the seemingly trackless 88 percent of Earth's history that constitutes the Precambrian. Yet in recent decades, the Precambrian has begun to yield its secrets.

We now know there were substantial organisms even before the Cambrian. They are known from some very strange but easily recognizable fossil animals, turned up in 1947 by geologists in the Ediacara Hills of southern Australia. The geological world was surprised to find out that these complex fossils came from a Precambrian layer and dated as far back as 700 My. These Precambrian lifeforms have come to be called Ediacarian life.

All of them were soft-bodied marine organisms, but they were firm enough and abundant enough to leave surviving impressions in ancient sedimentary rocks. One of the remarkable things about the earliest fossils is that they are not, as one might have guessed, very simple organisms, such as mere tubes to ingest waterborne microorganisms. Though soft-bodied, they appear to be complex creatures with well-formed systems of muscles, nerves, food-processing organs, and probably organs for sexual reproduction as well. Naturally, biologists would like to find good fossils of the "missing links" between them and their still earlier single-celled ancestors.

The most dramatic finds were animals that, strangely, did not seem specifically related to subsequent Cambrian organisms, which developed about 150 My later. As mentioned in Chapter 8, German paleontologist Adolf Seilacher tried to explain this by suggesting that a break in the chain of evolution occurred between the Precambrian and Cambrian, perhaps due to some eco-disaster. In shape, the Precambrian animals included oval, grooved creatures and circular flat disks with three bent arms spiraling out from their center, something like starfish. Others were somewhat more familiar. They included rounded, jellyfish-like creatures; frond-like animals resembling modern sea pens; worm-like animals; segmented animals with horseshoe-shaped heads, similar to the Cambrian trilobites that we will encounter in a moment.

Remember, too, that the period of the early Cambrian, just after 550 My ago, marks the appearance of still additional strange "alien" lifeforms, the so-called Burgess Shale lifeforms discussed in Chapter 8. Many of these Ediacarian and early Cambrian species seem to have been dead-end experiments in evolution—body types that did not survive. It was a time of great ferment in the long march of life.

THE EXPLOSION OF MARINE LIFE—700 TO 550 MY AGO

One can hardly overemphasize that the period of 700 to 550 My ago marks one of the most dramatic periods in the biological history of our planet because of the sudden expansion of lifeforms leading into the Cambrian Period, 550 My ago. This expansion had several important aspects. For one thing, hard-shelled creatures evolved and defined the beginning of the Cambrian. This alone guaranteed that fossils would be much more abundant than before, because the shells left prominent imprints in sediments instead of simply decaying. In fact, you might suspect that the abundance of Cambrian fossils was due to this aspect alone. But the Cambrian is more remarkable than that because of the rapid rate of the changes.

For example, the hard-shelled organisms appeared in quite a short interval—perhaps within 30 million years or less—in all the seas of Earth. They are found today in 550-My-old strata on every continent! Whatever happened to initiate the Cambrian, it allowed new lifeforms to spread with amazing rapidity. This is especially true of the two most representative Cambrian lifeforms: *brachiopods*, which look like small clams, and *trilobites*, which are oval or elongated animals with horseshoe-shaped heads and three-lobed bodies. Several species of brachiopods still exist today, but trilobites are extinct, being ancestors of modern crabs. Other species proliferated as well, including ancestors of snails, starfish, and sponges.

THEORIES TO EXPLAIN THE SUDDEN FLOWERING OF LIFE

The more geologists study the transition from Precambrian through Ediacarian lifeforms to the Cambrian Period, the more it seems that something special happened around 700 to 550 My ago. A reexamination of the Ediacarian animals helps

520 My ago. The Cambrian sea floor. The prominent trilobites were relatives of the distant ancestors of modern crabs, and reminiscent of the modern horseshoe crab. Their hard bodies and distinctive forms make them the most famous of Cambrian sea creatures.

450 My ago. Life begins the colonization of land. Early lichen-like plants (right foreground), hardy enough to survive among open-air rocks and soils, have spread from the coastal intertidal zones and are beginning to establish themselves independent of fluid water.

confirm this. Studies show that soft-bodied lifeforms, like the worms of the Ediacarian, tend to leave recognizable burrows and holes, even when their fossils are not well preserved. But no such remnants of these animals are common before about 700 My ago. Somehow, organisms became more complex at that time, evolved into shell-bearing organisms, and spread rapidly through all the seas.

Some geologists regard the start of the Cambrian—550 My ago—as the unique "moment" of change. Others would say that it was the whole 150-My period from 700 to 550 My ago. But the real question is why the changes occurred. At least five theories have been proposed.

Theory 1: Classic Darwinian Evolution. The first theory merely invokes classical Darwinian evolution, which says that competition and natural selection among species were enough to explain the rapid proliferation and spread. For example, the soft-bodied Ediacarian lifeforms proliferating through the seas were someone's lunch waiting to happen. The first predators could have had a field day with them, munching their way through the biosphere. The somewhat more hard-bodied forms, whether they evolved slowly or through individual mutations, would have been more likely to survive and have offspring. As further mutations occurred, the more armored ones preferentially survived.

430 My ago. Occasional episodes of glaciation are found throughout the geological record. Here runoff from melting glaciers carves a broad river valley. Lichens and other primitive plants give touches of color to continental landscapes, which are still unoccupied by substantial animal and plant life.

Brachiopods, trilobites, and other species gradually evolved. And no sooner did they appear than they found themselves so successful that within a few million years they could crowd out their competitors.

This theory can be criticized, however. Classical Darwinism was originally developed to explain a slow, steady evolution of life. To some biologists, the relatively sudden changes among many species all around the world, especially those around 550 My ago, remain puzzling. Some other influence seems called for.

Theory 2: The "Invention" of Sex. A second explanation for the flowering of life, either at 550 My ago or during the whole period 700 to 550 My ago, was the invention of sex. Curiously enough, the evolution of sex has long puzzled biologists. Even today, biologists ask (at least during their professional hours), "Why sex?" There are other ways to reproduce. For example, a bud containing the organism's DNA and genetic information can develop and break off from the parent to form a new individual. Some bacteria and simple plants reproduce in this way. Other bacteria sometimes conjugate to exchange genetic material, in a process resembling simple sexual activity.

The answer usually given to "Why sex?" is that sexual reproduction took precedence because it allowed rapid mixing of the genetic material of a species, which in turn caused "good" mutations to propagate rapidly. This could explain why a trilobite or a brachiopod, once evolved, could spread rapidly around the world. The idea is that sexual reproduction evolved among the very simple Precambrian organisms shortly after 1000 My ago, and that once this form of reproduction developed, it greatly accelerated the rate of evolution and the spread of distinct species with complex sexual behavior.

Some biologists are uncomfortable with this explanation of the supposed reasons for sex. The argument is vague. To the extent that science can ever answer "why" questions, the answer to the question "Why sex?" may have more to do with the ultimate

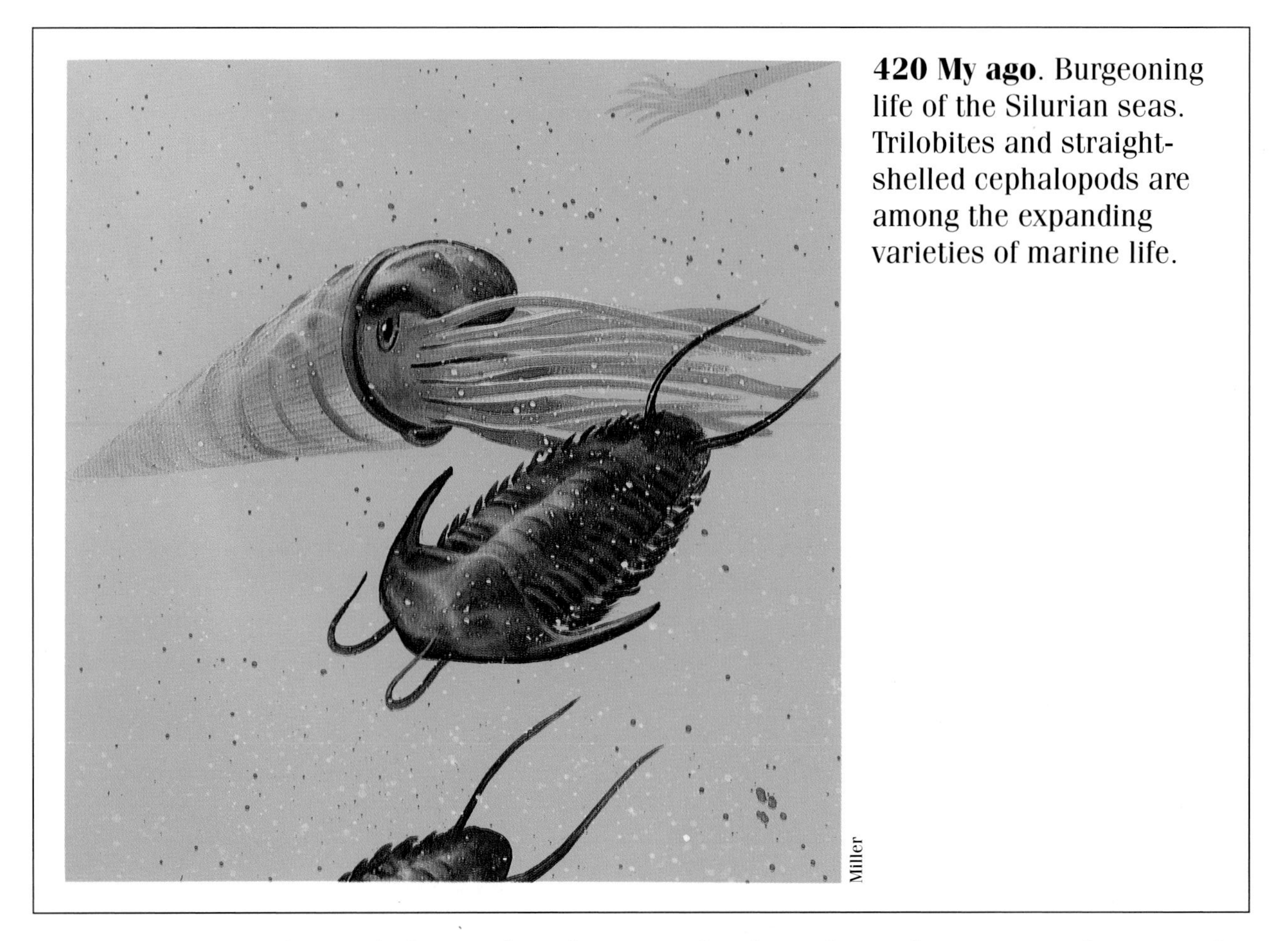

420 My ago. Burgeoning life of the Silurian seas. Trilobites and straight-shelled cephalopods are among the expanding varieties of marine life.

molecular nature of life. Let's look at sex from that perspective.

At the molecular level, as well as with a courting couple, two is a magic number for life, because the complex organic molecules that originated life and carry genetic information split into two halves, and then each half can reconstitute itself into a new individual. According to our molecular viewpoint of life's origins, life is a process that reduplicates, and hence multiplies, very complicated organic molecules. In nonsexual reproduction of such molecules, the half molecules have to go off and assemble new second halves one atom at a time, or one cluster of atoms at a time, from the surrounding seawater. But in sexual reproduction, the half molecule is not just sent out to fend for itself; it is encased in its own cell, called a gamete. Unlike most cells, which have complete double-organic molecules with complete genetic information, the gamete is only half complete and finds another gamete with another half set of genetic material. The union of the two gametes makes a new whole cell all at once.

Evolution produced multicelled organisms, but always a goal of the increased complexity was to protect the genetic material. In multicelled organisms, the new whole cell was a fertilized egg. It is interesting to note that in many seafaring species, the fertilization takes place outside the body: the gametes are simply sprayed out into the water and the union takes place there. This is a powerful echo of the early history of life in Earth's seawater. The gametes drifting through the water mimic the earliest simple organic molecules and their descendants, the earliest lifeforms. In the same vein, we are familiar with male fishes fertilizing eggs laid by female fishes. The new whole cell—the fertilized egg—ultimately grows on its own until it produces a new set of gametes.

It is from this point of view that advanced organisms, especially land animals, are giant machines designed to nurture their own gametes and their offspring. Sexual reproduction works by creating, one might say, half organisms: female and male. Each grows until it can generate well-protected gametes that have only half the normal genetic material: eggs in the female and sperm in the male. Of course, the fertilized egg is a whole organism, and most of the cells of our body are whole in the sense of having the complete set of genetic material. But their jobs, from a molecular point of view, are simply to produce and protect gametes—cells that are only half complete but can combine with other gametes to make new organisms. Sex thus saves the laborious process of reproducing DNA, RNA, and other genetic molecules one atom at a time in the wild. Sex may be explained not so much by its evolutionary consequences, which came after the fact, but by the very nature of the most complex organic molecules, with their two separable halves, in our deepest being—the nuclei of our cells.

In this sense we males and females really are each other's long-lost other halves. The sense of wholeness in our companionship and our sexuality, celebrated by romantic poets, is in this context enmeshed in the very nature of our organic molecules.

We tend to view sex from a land-dwelling mammal's viewpoint, giving great emphasis to intercourse between males and females, not to mention the ensuing growth of the baby inside the mother's protective womb and the transitional stage of birth. But from the molecular point of view, these are merely embellishments that allowed life to adapt to the harsher, alien environment of land—an environment that life had not yet even adopted 550 My ago.

By reproducing a whole cell at a time, sex may have been a more advantageous way to spread life, and certainly a more interesting way from our point of view, but no one has been able to prove that it caused the proliferation of life 700 to 550 My ago.

Theories 3 to 5: Environmental Changes. A third theory has been that the proliferation of life was caused by the dramatic change in the atmosphere mentioned earlier—the rising oxygen content. Certainly the conversion to oxygen would have required dramatic changes in lifeforms and would have facilitated the rise of animals. Further, the potent chemical activity of oxygen guaranteed efficient production of energy in oxygen-breathing species—consistent with the vigorous spread of new animals around the world.

A fourth theory takes note of the evidence about early glaciation. If the climate was colder and glaciation extensive from 950 to 650 My ago, then the climate and the seas probably warmed after 650 My ago. This more clement condition might explain a sudden proliferation of life by 550 My ago.

There are problems with attributing the great flowering of life to oxygen or to spontaneous climatic warming. One problem is that the rise of oxygen content was an extended process starting 2000 My ago, not something that happened 700 to 550 My ago. Similarly, the cause of a sudden spontaneous warming is unexplained. What seems to be needed is some cause that would create an abrupt change.

For this reason, a fifth theory, a new type of catastrophist theory, has been suggested in recent years involving some extraterrestrial influence that caused a sudden environmental change. As we will see in Chapter 13, strong evidence suggests that impacts of asteroids have altered the course of biological history. Perhaps one such large impact altered climatic or oceanic conditions around 550 My ago in a way that cleared environmental niches for the rapid emergence of new species. We know from dates of ancient, eroded craters that large impacts have occurred sporadically during the last few hundreds of millions of years, but connecting them with specific evolutionary events, such as the flowering of life at the start of the Cambrian Period, is difficult.

360 My ago. In the late Devonian swamplands. Plant life has surged onto the land, and one of the early amphibians, about a meter long, ventures out of the water into the still-unfamiliar but uncontested environment above the waterline.

For these reasons controversy remains about why Earth experienced such an explosion of lifeforms. Geologists and paleontologists will continue combing Earth for clues as to whether it was due to normal evolutionary processes, sex, oxygen, climatic warming, or asteroidal impact.

WHAT FISHES TELL US

The early geologists who identified the Cambrian Period from fossils also defined several succeeding periods based on changes in characteristic fossils as time went on. Life was now evolving so fast that these geologic periods were less than 100 My long. The Ordovician Period started at the end of the Cambrian, 500 My ago, and ran until about 440 My ago. The Silurian Period followed, from 440 to 405 My ago. Then came the Devonian, running from about 405 till 360 My ago.

The creepy trilobites and quiet, clam-like brachiopods of the Cambrian seas were only a harbinger of things to come in the seas. By 450 My ago, not only shelled sea creatures were proliferating but also fishes. First came jawless fishes who swam along the bottom, shoveling up organic-rich sediments from the sea floors. By the early Devonian, about 400 My ago, these were being replaced by more familiar fishes with jaws. But what fearsome beasts! There were shark-like carnivorous fishes thirty feet long, and fishes with massive bony armor plates for protection from the carnivorous predators.

Because of its abundance of fossils of awesome fish and the relative absence of land animals, the Devonian Period is often called the Age of Fish, and the proliferation of the fish reveals to us the explosive rate of evolutionary change once advanced life got started.

PLANTS SURGE ONTO THE LAND—400 MY AGO

These were the periods of still another revolution in Earth's history. Lifeforms came out of the sea and colonized the land. Primarily, these were plants. As recently as the 460 My ago (90 percent of the way through Earth's history), evidence of life on land was hard to find. Paleontologist Jane Gray and her colleagues at the University of Oregon have reported ancient spores indicating the first presence of simple land plants at that time. Mosses, lichens, and other primitive plants may have grown here and there on the land by 450 My ago. All such plants left few fossils because of their insubstantial nature. A few fossils of poorly developed plants may date from this period. But within only 50 My, by 400 My ago, the land was teeming with strange forests and the first land animals. It took plants only 1 percent of Earth's history to transform the continents. Not so hard to believe when you think about what weeds do to your garden in one summer!

Blasé though we may be about living on land, the conquest of the land was an extraordinary transition for life. Life is an aqueous enterprise. In a sense, we are still sea creatures. We carry our fluids inside us, and our skin is a coating designed to keep our insides from drying out. A provocative fact is that our bodily fluids and those of our fellow animals have almost the same proportions of various salts as seawater, a clear indication of our legacy from the sea.

Life on land requires protection from a dangerous environment. Had Cambrian sea creatures been intelligent, the conquest of the land would have seemed as dangerous to them as the conquest of space does to us. The dangers were several. First there was sunlight. The sun, great nurturer of life, is also a great threat. Its ultraviolet light breaks down organic molecules and causes skin cancer. Plants and animals needed to evolve protective coverings.

Second, the dry air desiccates moist cells. In particular, reproduction was threatened. No longer could organisms simply eject their gametes, or sex cells—eggs and semen—into the environment outside the body and hope for the best. One of the big problems for lifeforms on land was to protect the gamete cells. The solution to this problem for plants was the evolution of seeds in protective husks; for animals it was fertilizing the egg *within* the body of the mother, giving it a sea-like environment during gestation.

A third problem was the maintenance of the salts in body fluids. Plants and animals probably colonized the land by making the intermediate step of evolving to survive in freshwater surroundings, free of the salts of the ocean water. But by ingesting salt-free fresh water on land, they endangered the sea-like conditions of their inner fluids. New mechanisms had to evolve to maintain the salt ion balance. Animal life in particular can be viewed as big, wandering packages containing sea-like environments. Recognition of the need to maintain the salt ion balance in our interiors explains the sudden international popularity of ion replacement beverages, such as Gatorade. The adjustments to land started with plants but had to come step by step. Green algae seem to be the ancestors of chlorophyll-bearing land plants, such as ferns, conifers, and flowering plants. Green algae today flourish in fresh water, an example being the green scum often found in stagnant, landlocked ponds. Another example of early land plants include mosses and liverworts. A clue to their primitive nature is that they lack efficient systems for transporting fluids throughout their bodies, and they also lack adequate means to keep their gamete cells from drying out when exposed to the sun. For these reasons, even today they have failed to make a complete transition to land: they flourish only in shady, moist environs. Lichens give the perfect demonstration of the first steps in colonizing the land: the sea-loving algae could not have colonized the sun-blasted land without the shade and protection of the fungi.

Lichens are another candidate for very primitive land colonists. They are symbiotic combinations of fungi, which cannot produce their own food, and green algae, which conduct photosynthesis. The fungi protect the algae from the harshness of sunlight and dry air, and also allow the lichen to anchor to rocks and other surfaces.

Despite the problems, land offered a fantastic, unexploited environmental niche for early plants. Those that grew on riverbanks had a convenient supply of water but were out of the reach of foraging fish. Minerals were available in moist soils. The now-oxygenated air offered direct availability of oxygen to support animal life. Sunlight, a source of danger, was also a direct source of energy for photosynthesis. Once plants colonized the land, the land became an extraordinary environmental niche for animals, because the soil-rooted plants themselves offered food for the taking. There was no competition for the first animals capable of making forays onto the land; the land was unoccupied. Thus only in strata laid down after plants became common on land do we find abundant animal fossils.

THE FIRST FORESTS

While lacking large animals, the landscapes of the Devonian saw incredible changes in only about 50 My. Plant life greened once-barren land for the first time.

Any plant whose structure promoted capillary attraction or other means to pull moisture and minerals from the moist soil could survive inland, away from rivers. Plants with root-like systems were thus favored, and the root systems evolved rapidly. Also quick to appear were so-called vascular systems: systems of cells that transmit moisture up the stalk of the plant. The more surface area exposed to the sun, the more photosynthesis occurred. Evolution favored specialized organs to expose surface area to the sun: leaves. Plants need sunlight. The taller the plant, the better the chance of getting out of the shade of other plants. Taller plants needed stronger stems. The earliest stiff-stalked and woody plants appeared around 400 My ago. Stiffness enabled plants to grow vertically and stand taller in the air—an advantage as plants proliferated in the vast unexplored "environmental niche" that was the surface of the land.

Many Devonian plants were seedless. Ferns are one type of plant that has survived since the Devonian. Most others had a less leafy appearance. Ferns bear interesting traces of their origins. They produce no seeds but rather spore cells that fall to the soil and grow into a peculiar intergeneration plant that is not a leafy fern but a specialized fungus-like form that produces egg cells and sperm cells. Propelled by its whip-like tail, the sperm must swim a short distance through the water of moist soil to fertilize an egg cell from a neighboring plant. The fertilized egg then grows into a mature fern that produces more spores (produced in the brownish spots on the underside of fern leaves). It is as if the fern is an amphibian plant, growing in the air but reverting to water to transmit its gametes.

Most of the seedless plants that flourished in the Devonian forests of 350 My ago have died out, although ferns are common in many of our woodlands. The giant tree fern, another descendant of the plants of this era, is still common in tropical regions. For example, you can walk through forests of tree ferns in the prehistoric-looking landscapes of Hawaii Volcanoes National Park.

REEMPHASIZING THE SUDDENNESS OF LIFE'S PROLIFERATION

It is hard to overemphasize the shocking suddenness of life's proliferation in the seas and on the lands of Earth. Suppose some galactic civilization had evolved on planets of stars older than the sun, and they had sent out a dozen expeditions to visit Earth, spaced evenly throughout Earth's history. The first ten expeditions would have found only lifeless craters, lavas, sand dunes, and highly eroded river channels on Earth's land surfaces. The eleventh expedition would have arrived 380 My ago and would have found the land mostly covered by flourishing Devonian forests. The twelfth expedition would find *us*.

400 to 100 My Ago

11. The Continents Take Shape

> If we are to believe Wegener's hypothesis, we must forget everything that has been learned in the past seventy years and start all over again.
>
> *R. T. Chamberlin, during a meeting of the American Association of Petroleum Geologists in 1926*

Several times we have mentioned that the Earth's crust is mobile, and that the continents have changed positions and shape. This was a hard concept for geologists to accept. Agreement on this idea came only as recently as the late 1960s, and it caused a revolution in geology. Modern geological interpretations of Earth's surface would be impossible without this discovery.

PLATE TECTONICS

We now know that the crust is divided into mobile sections called plates. Each plate can be thought of as a rock sheet, riding atop the partly molten zone. Currents in the molten material help drag the overlying plate in one direction or another. Although we think of rock as rigid, it is actually plastic and deforms by flow if given enough time. That is one of the secrets of the plates. When two plates collide, instead of simply shattering, the layers of rock may be arched, like a deck of cards being compressed from both sides.

The surface of a plate is generally being carried in one direction—for example, from south to north. Usually, lava wells up at one or more edges of a plate, because of ascending currents from the mantle. Lava thus emerges from the mantle on that side of the plate, helping to push the plate away from that zone. The far edge of the plate may collide with another massive plate and be crumpled.

Some plates contain continents, or large parts of them, so that the continents themselves drift as the plates move. In some cases, an oceanic plate of dense lava collides with a continental plate composed of lighter granitic rock. The lighter rock tends to float on the denser rock, so it may ride up over the approaching oceanic plate. The edge of the oceanic plate sinks into the mantle under the continent's edge, eventually melting and being mixed with upper mantle material.

The study of the motions of large regions of the crust is called *tectonics.* Although the theory of mobile continents was originally called the theory of *continental drift*, it is now called the theory of *plate tectonics* because the tectonic motions involve not only continents but also other plate surfaces, such as sea floors. From its effects on continents, plate tectonics might be called rifting and drifting: pieces of continents tend to break off along fractures known as rifts, and the pieces drift and merge with other continents.

When we hear Shakespeare's line about all the world being a stage, it still seems subconsciously evident that Earth's surface is as unmalleable as a wooden platform. Of course, we've known about earthquakes and volcanoes since Shakespeare's day, but over the long term, land surfaces seemed fixed. Only a few people prior to the last few decades talked seriously about whole continents moving relative to one another.

The first publication to discuss the idea was a 1858 religious book by one Antonio Snider-Pellegrini: starting with the account in Genesis, he enumerated events between the Creation and the Deluge. The book was hardly scientific, but it did point out the curious fit between the coastal outline of South America and Africa, on opposite sides of the Atlantic. Snider-Pellegrini suggested that a giant proto-continent split apart during the Deluge on the sixth day of creation, forming the Atlantic. This jigsaw fit of the two sides of the Atlantic became a mainstay of continental drift theorists, even though the rest of Snider-Pellegrini's arguments sank into oblivion faster than a dense blob of crustal rock.

A more scientific approach came in the first geophysics text, *Physics of the Earth's Crust*, published by the British clergyman and naturalist Osmond Fisher in 1881, and in a 1910 paper by an American geographer-geologist, F. B. Taylor. Fisher and Taylor based their conclusions not so much on the Atlantic as on more abstruse geophysical arguments. They criticized the then-current idea that folded mountains (mountains with crumpled strata, like the Alps or Himalayas) were caused by shrinkage of Earth during cooling (like wrinkling of an apple's skin). Instead, they argued—correctly, as it turned out—that Earth's interior was fluid, and that continent-sized blocks of crust floated and moved on this interior. Fisher argued that hot currents ascended in the mantle under the oceans, widening the Atlantic, for example. Taylor argued (also correctly) that India had collided with Asia, crumpling the contact zone to form the Himalayas. The work of these two had little influence at the time because the idea of moving continents seemed too fantastic.*

*The history of continental drift and plate tectonic theories is discussed further in the illuminating book *Great Geological Controversies* by A. Hallam.

ALFRED WEGENER: A VOICE IN THE WILDERNESS

The first geologist to attract wide attention to this theory was also the first to discuss detailed physical evidence and also possible mechanisms. He was the German naturalist Alfred Wegener (1880–1930). As a young man, Wegener conceived a passion to explore Greenland, and in training himself for this mission, he established a record for the longest balloon flight—fifty-two hours—in 1906. His doctoral degree was in astronomy, but he also became interested in meteorology and geophysics, the focus of his eventual university position.

Wegener said that he got the basic idea in 1910 from the fit of the Atlantic coastlines; there is also some evidence that he was influenced by the splitting and separating of glacial ice blocks he saw in Greenland. At any rate, he was spurred on by an important additional piece of evidence in 1911, when he read of similar fossil types found in both Brazil and Africa—a strange finding given that South America and Africa are notable today for different species, such as the New World monkeys and Old World monkeys. To explain the fossils, most geologists of Wegener's day speculated that there might have been a prehistoric land bridge across the ocean between the two continents. Wegener, however, correctly proposed the idea that the continents had once been a single mass, with a single community of plants and animals, and that the continent later split along the Atlantic, allowing more recent species to evolve independently in two separate environments.

Wegener lectured on these ideas as early as 1912, and promoted them in a book, *The Origin of Continents and Oceans,* published in several editions in the 1920s. Here he used some geophysical arguments that remain valid today. First, he noted that Earth could not have contracted enough to explain the massive crumpling of mountains like the Alps, where initially horizontal layers have been

Hartmann photo

Faulting dropped the foreground valley and raised the background mountains. Subsequent erosion carved the mountains into steep peaks and deep valleys. Grand Teton National Park, Wyoming.

folded and torn apart over distances of many miles. Second, he noted that measurements of rock types and of the force of gravity over oceans and continents showed that the ocean basins have high-density rocks, while continents are made of a different rock type, more granitic and lower in density. Wegener argued (again correctly) that the low-density continents float on the higher-density material, like solid scum on the surface of molten slag. This shows not only that continents are distinct rock units, but also that the ocean basins could not be formed simply by vertical sinking of old continents, as most geologists then thought. Third, Wegener not only found similarities of fossils on the coasts on opposite sides of the Atlantic, but also found matches of rock types on the two coasts—which he likened to finding two pieces of torn newspaper not only with a matching shape of tear but also with matching lines of type. If the lines of type matched, he said, the two parts of the paper must once have been one.

Applying the fossil chronology, Wegener postulated (again correctly, as it turns out) that these landmasses were joined together into one huge continent during the Devonian Period and other periods shortly thereafter—the periods we discussed in the last chapter, about 300 My ago. Wegener called this giant landmass Pangaea, or "universal land."

The big problem with Wegener's argument was that no one could figure out what force could possibly drive solid rock continents across Earth's face. What could push rock ponderously across other rock? Wegener may be forgiven his lack of an answer; it took years of geophysical work to solve the puzzle. But from the 1920s to the 1950s, this problem seemed a fatal flaw. Leading geologists and geophysicists roundly criticized Wegener and his theories. Sir Harold Jeffreys, one of the greatest mathematical geophysicists of this period, made calculations to show that the continents couldn't move through the supposedly stiff, resistive medium in which their undersides rested. Several editions of his book *The Earth* blasted Wegener's ideas, and by the time I was a student in geology graduate school during the early sixties, Wegener had been relegated almost to the status of a crank, and continental drift was on the fringe of legitimate geology.

HOW MAGNETISM HELPED ESTABLISH PLATE TECTONICS

In the 1950s, the new geological technique of paleomagnetic measurements emerged, as described in the last chapter. When applied to continents and sea floors, it had shocking results.

As we already saw in Chapter 10, magnetism showed that continents had been at different latitudes during different geologic periods. This confirmed Wegener's idea of continental movement.

This conclusion began to be heard in the early 1960s from S. Keith Runcorn and his associates at Cambridge, England, and other researchers. Yet even in the mid-sixties many classical field geologists remained dubious. I recall a seminar that Runcorn gave when he visited our university; afterwards, my geology professors shook their heads sadly and opined that magnetic measurements were extremely delicate and unreliable; and after all, how could the continents move?

But Runcorn and Wegener turned out to be right.

WHAT SEA-FLOOR ROCKS TELL US

The breakthrough evidence for plate tectonics came when magnetic measurements were applied to sea-floor rocks. A strange property of Earth's magnetic field was involved. Every so often it reverses polarity, so that a compass points south instead of north. In the last few million years, this has happened every few hundred thousand years, but in earlier periods the reversal rate has, at least sometimes, been slower. The cause of these changes is one of the mysteries of geophysics.

Thus adjacent rocks formed during different magnetic eras can have reversed polarity of their magnetic domains, relative to each other. On the sea floor, this led to an amazing result. By the beginning of the 1960s, mapping of the sea floors had revealed huge systems of ridges, which were the source of lavas. A famous example is the Mid-Atlantic Ridge, which winds midway down the center of the Atlantic basin, paralleling the American and Euro-African coastlines. During the early sixties, sampling and dating of sea-floor rocks revealed that the rocks at the ridge were very fresh, but that the farther one traveled from the ridge, the older the rocks. Generally, all the Atlantic floor rocks were surprisingly young, a few tens of millions of years old.

Fresh magma rises from sources in the mantle under the ridge and erupts at the ridge crest. The exciting aspect of the sea-floor data was the discovery that alternating bands of reversed paleomagnetism exist at different distances from the midocean ridges. They erupted at different times in the past, when Earth's magnetic polarity was first in one direction and then in the other. These results on both dates and magnetism confirm that the lavas erupt at the ridges, but then are slowly pushed outward from either side of the ridge.

Oldest lavas are farthest from the source along the ridge crest. Two zones of lava that formed by flowing onto either side of the ridge 50 My ago, as identified by their dates and magnetic polarity, might now be on opposite sides of the Atlantic sea floor. In other words, the whole sea floor has been spreading apart, confirming Wegener's ideas once and for all.

It was hard to establish these new ideas and overturn the old idea of fixed continents. Many of the papers of the early sixties were considered by the geological establishment to be wild speculation, despite the new data from the sea floor. It was hard for geologists to accept the new ideas. In one now-famous case, Lawrence Morley submitted a letter to the journal *Nature* in 1963, discussing "the concept of mantle convection currents rising under ocean ridges, traveling horizontally under the ocean floor and sinking at ocean troughs" with magnetic signatures recording the alternately reversing magnetic field. The ideas are now known to be correct, but the letter was rejected and not published. Papers submitted to such journals are always sent to other scientists for reviews, usually on an anonymous basis to help the editor decide whether to publish the material. As quoted by the historian A. Hallam, a reviewer in this case dismissed Morley's letter with the now-notorious comment, "This is the sort of thing you would talk about at a cocktail party, but you would not write a letter on it."

Despite such naysaying, the geological community was almost entirely convinced of continental drift by the late sixties. A few crucial meetings, in which all views pro and con were aired, helped establish the new ideas. The Canadian geophysicist J. T. Wilson was first to use the term *plates* in 1965. The theoretical framework quickly became known as the theory of plate tectonics. By 1970, virtually all geologists believed in plate tectonics and realized that it was the key to further progress in understanding Earth.

BASIC IDEAS OF PLATE TECTONIC THEORY

The key in overturning the old ideas was that much of the upper mantle is not rigid, solid rock but plastic, deformable rock near the melting temperature. The earlier geologists who visualized the upper mantle as solid rock could not see how crustal plates could plow through it; but the new ideas made it easier to visualize sluggish mantle currents in partly molten material, carrying the thin, more rigid crustal plates along. The whole upper mantle can flow, given enough time.

This most plastic zone, about 100 kilometers down, is called the *asthenosphere* (from Greek roots meaning "weak layer"). Its plastic properties have been confirmed by the study of earthquake waves that travel through this region. A key concept is that although rocks may seem rigid to us, they act more like fluid over long geological time periods. This is why some cliffsides and road cuts expose strata that have been gracefully bent in huge arches; the rocks did not shatter, but deformed like plastic. The solid rocks in the asthenosphere would seem rigid to us if we could examine them. But they can flow and deform on long time scales. Such behavior is really not so unfamiliar. An example is asphalt or tar: you can shatter a piece of it with a hammer, but if you leave a heavy weight on it long enough, it will flow and form an imprint of the object. In the same way, materials of the asthenosphere flow over long time scales of millions of years. Such behavior is often aided by heat: the tar is brittle on a winter day but deforms easily on a summer day. Due to the heat coming out of the hot interior of Earth, the material of the asthenosphere is believed to follow great circulation patterns called convection cells.

This phenomenon can be seen in an experiment in your kitchen. Place a pan of cooking oil on the stove and heat it. (Be careful—the oil can get much hotter than boiling water and causes severe burns if spilled.) Under a strong light, you can see convection patterns where hot oil rises in some areas, flows across surface regions, and descends elsewhere. This divides the surface region into "plates," separated by the rising "hot spots" and the plunging regions. Sprinkling a scum of flour on the surface sometimes helps to define the plates.

The outer 70 kilometers or so of Earth, including the crust and some mantle rocks just below the crust, are cooler and stiffer than the underlying asthenosphere. This is the *lithosphere,* or "rock shell." Although the lithosphere can deform plastically, it is rigid enough that it sometimes cracks, forming boundaries of plates. New lava may erupt along such fractures, as at the Mid-Atlantic Ridge.

The plates float on the more plastic, moving layer and are dragged along with it. Depending on the mantle flow patterns during any given geological era, one plate bearing a continent may carry it toward the edge of another continent. The collision of two continental blocks may last millions of years. The lithosphere rocks are slowly bent and crumpled, creating a mountain chain along the continents' margin. The Himalayas are the best example: they formed when India crashed into southern Asia.

A similar crumpling can occur even without a collision of two continents. If a continent is pushed too fast by its plate into an adjoining sea-floor plate, its leading edge crumples and makes a mountain

350 My ago. Endless unspoiled beaches. In the world of the Carboniferous Period, coastlines included hundreds of miles of a vacationer's dream.

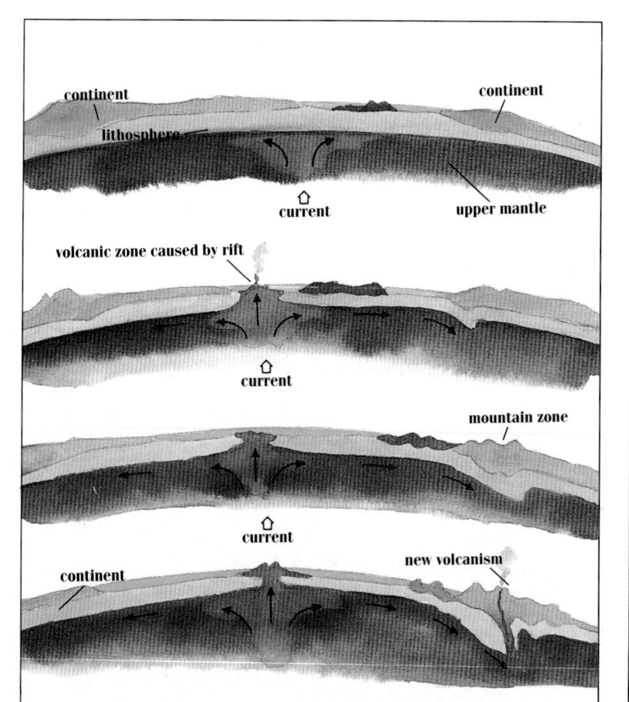

Some principles of plate tectonics, as illustrated by a cross section of Earth's crust and upper mantle. Slow currents welling up from the mantle cause rifting of a section of crust, which creates a volcanic zone and forms a plate boundary. In this example, the sea-floor crust on the right collides with a continent and is dragged under it, crumpling the margin of the continent and creating a mountain zone. As the crustal rock is dragged under this zone, parts of it melt and new volcanism is started along the mountain zone, which might resemble the Cascade range or the Andes. Volcanism continues along the original rift, which now resembles the Mid-Atlantic Ridge system.

chain, usually arc-shaped. This explains some mountain zones at the margins of some continents, such as the Andes of South America. At such collision zones, where dense sea-floor lithosphere hits lower-density continental lithosphere, the sea-floor plate may slide under the edge of the continent and sink into the mantle. As part of one plate is dragged downward under the other, it creates a depressed trough adjacent to the crumpled mountain zone. This explains the deep ocean troughs off the shores of some continents.

Of course, humans don't feel the motion of plates, or of the continents as they are dragged along with the plates. All the motions are too ponderous. It is as if we were ants on the surfaces of icebergs. The continents seem stable and everlasting. Even the mountain-building crumplings are processes that extend through geologic eras, raising ranges and tearing troughs little by little. What we experience is only the occasional earthquake, when underground rock layers snap under the pressure, or volcanic eruptions, when hot magmas from the upper mantle break through to the surface.

If a new major hot spot develops under a continent, due to magmas ascending from the mantle in a convection current, it will cause tension in the overlying continental plate. Eventually a fracture may split the continent in a long, depressed fissure called a rift. The two sides of the rift are pulled apart by the ascending, spreading magma. Volcanoes erupt along the rift because magmas gain easy surface access through the fracture zone. Lava pours out, filling the floor of the rift. Since the rift is depressed, it is likely to be filled with water. As the two sides of the rift move apart, this sea widens, and the volcanic fracture along the middle becomes a mid-ocean sea-floor ridge. This is what happened around 150 My ago when the Americas pulled apart from Europe and Africa, creating the Atlantic and the Mid-Atlantic Ridge.

UNFAMILIAR CONTINENTS OF THE PAST—2000 TO 700 MY AGO

One of the exciting things about plate tectonic theory, aside from its explaining many of the *processes* of Earth's surface evolution and structure, is that it allows us to reconstruct, at least in part, how ancient Earth looked. Apparently the continents have been drifting more or less at random ever since continental crust first appeared on Earth. Like masses of ice in a river or islands of soapsuds in a bathtub, they were sometimes well separated, but sometimes jammed together. By assembling measurements of modern plate movement rates, fossils, and geologic structural data from sites at known dates in the past, along with paleomagnetic data, we can reassemble the patterns of ancient continents partway back in Earth's history.

The farther back in time we go, the harder it is to get enough data to locate the continents. Paleomagnetic data are poorer, and it is harder to find fossils, glacial deposits, or rock types that match up on, say, the coast of China and North America to show they were joined at that time. The most ancient period for which reconstructions have been attempted is the interval of 2500 to 1000 My ago. For example, a region in southern India has been identified as the site of a collision between two plates 2500 My ago.

Maps from different experts do not agree very well for this period. There is some indication that the continents were rather spread out around 2500 to 2000 My ago. According to some data, by 1700 to 1500 My ago, southwestern Asia, North America, South America, Africa, and Antarctica were a more or less continuous landmass. Much of this occupied the Northern Hemisphere, with the North Pole being located near the Asia-America contact, perhaps near the region of present-day California, and Africa lying along the equator. By 1000 My ago, however, much of this landmass had shifted toward the equator, with Africa slipping into the Southern Hemisphere

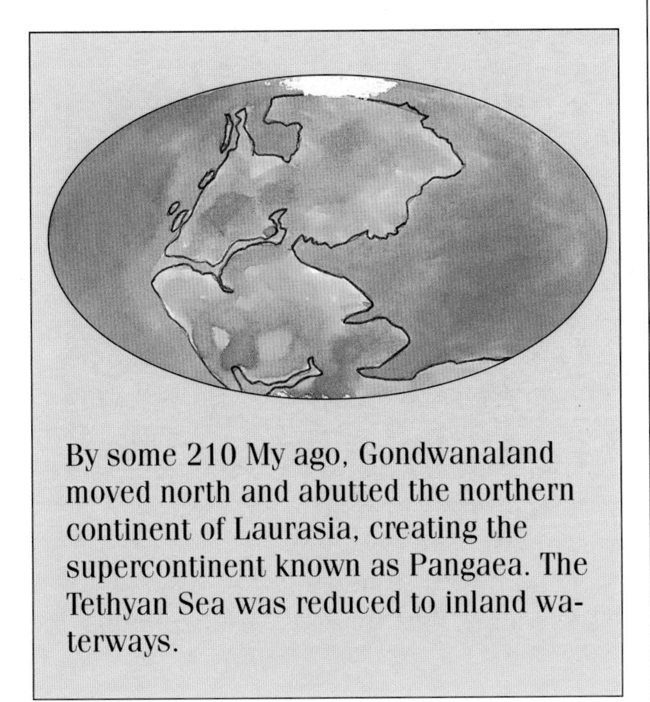

By some 210 My ago, Gondwanaland moved north and abutted the northern continent of Laurasia, creating the supercontinent known as Pangaea. The Tethyan Sea was reduced to inland waterways.

GONDWANALAND AND LAURASIA—500 TO 400 MY AGO

By the era covered in the last chapter, different authors' maps of continental distributions begin to agree and become more reliable. By 500 My ago, a supercontinent existed in the Southern Hemisphere. Africa and South America were firmly joined; India was part of eastern Africa (joined just "below" present Arabia), and Madagascar and Antarctica were melded onto the present southwestern coast of Africa. Continental drift carried what are now parts of Africa over the South Pole some 500 to 400 My ago. The northern and later southern parts of that continent were covered by the South Polar ice cap, explaining massive glacial deposits in present-day desert and tropical regions, as well as in adjoining South America during this period.

The Southern Hemisphere supercontinent of that period is called *Gondwanaland.* It was named by one of the earliest theorists of continental drift, the Austrian geologist Eduard Suess, in 1885. The name came from a region of eastern India, Gondwana, known for fossils that were also found in Africa. As early as the era of Suess and Wegener, evidence had emerged that these landmasses were much closer together than at present, and perhaps at a more southern latitude.

320 My ago. The scenery typical of high mountain country is primarily a product of continental drift, because high mountain ranges are produced during collisions of tectonic plates, when large masses of strata are thrust to high altitude and then eroded into sharp peaks by running water. Some of the high coniferous forests were beginning to look like those of today.

400 My 100 My

To the north of Gondwanaland, along the equator and at low northern latitudes, lay North America, Europe, and Greenland. This explains why these areas have ancient deposits that originated under desert or tropical conditions, such as red desert strata, salt deposits left by evaporation, and coral reefs, dated about 600 to 400 My ago. Europe and America, in modern times linked so closely culturally, were at that time physically united along what is now eastern Canada. Asia lay slightly farther north and east, in mid-northern latitudes. During this interval about 400 My ago, North America, Europe, Greenland, and Asia aggregated into one fairly coherent supercontinent, broken by shallow waterways and seas. This supercontinent has been called *Laurasia* (the first half of the name coming from so-called Laurentian formations around the St. Lawrence River in Canada).

In the early literature of continental drift, there was some argument as to whether modern continents came from the fragments of two large masses, Gondwanaland and Laurasia, or from fragments of one giant landmass, which Wegener had called Pangaea. Today we see this less as an either-or choice than as different stages of development. According to modern usage, Pangaea was the next stage.

PANGAEA—300 TO 200 MY AGO

A crucial aspect of Earth's history around 300 My ago was that Gondwanaland and Laurasia collided to form Pangaea. Gondwanaland moved north, shifting the glaciation in Africa to what is now Africa's southern tip. A narrower and narrower waterway, sometimes called the *Tethyan Sea*, separated Laurasia on the north from Gondwanaland on the south.

Eventually Laurasia and Gondwanaland merged loosely along the equator. The present east coast of North America, Venezuela, the Sahara region, the Mediterranean region, and Arabia all nestled together, probably with patchy remnants of the Tethyan waterways among them. The present Mediterranean is one of these surviving relics of the ancient Tethyan Sea.

For this brief period, most of Earth's landmass was concentrated in one super-duper continent, Pangaea. This was just a random configuration of a random moment in Earth's history, but it has had interesting ramifications for life on our planet. By chance, it happened late enough that some of the most striking lifeforms, such as the dinosaurs and giant trees of the Carboniferous Period, had evolved, but also early enough that modern lifeforms had not evolved. Given one giant continent at clement latitudes, ancient species of plants and animals were able to spread over most of the land surface of Earth. This is the reason that fossils of many species of dinosaurs and other ancient land creatures are found on all the continents today, and that the geologists of two generations ago were so successful in mapping and dating strata confirming that Pangaea once existed. If Pangaea had formed and broken up only a few hundred million years earlier, there might not have been enough dramatic fossils to establish that it ever existed.

TOWARD MODERN CONTINENTS—200 TO 100 MY AGO

Pangaea did not last long. The Tethyan Sea once again began to widen, and Laurasia once again began to separate from Gondwanaland. Meanwhile, slow currents were rising in the upper mantle, the asthenosphere, under the region where Europe and North America were joined. A rift began to open in this region, and the North Atlantic Ocean began to form. This time Europe changed allegiance, staying with Asia instead of with North America. A new plate boundary was now establishing itself between Europe and America. Along the rift defining the new boundary, volcanism was widespread. South America was still joined to Africa until 100 My ago, when it too began to pull away, opening the South Atlantic.

Miller

90 My ago. Vast rain forests and jungles of enormous ferns and primitive palms cover much of the tropically climated Earth during the Jurassic Period—the heyday of the giant reptiles, when they reached the peak of their dominance.

In an unrelated development, Antarctica/Australia and an Australia-sized continent, India, rifted apart from eastern and southern Africa. Plate tectonic motions took Antarctica/Australia toward the South Pole, while India began its drift northward toward the southern coast of Asia.

From the point of view of an alien mapmaker passing by Earth in a spaceship, Earth was only beginning to take on its familiar modern appearance. The time was 100 My ago. North and South America were more or less recognizable, though flooded here and there with inland seas. The Atlantic Ocean separated the Americas from Europe and Africa. The brooding Eurasian continent dominated the Northern Hemisphere. Half of Earth was taken up by the vast Pacific. The Tethyan Sea had become unrecognizable, replaced piecemeal by the Mediterranean, the Black Sea, and the Indian Ocean. Antarctica was moving toward the South Pole. Australia was breaking off from Antarctica, and India was moving north toward the collision that would join it to Asia and raise the mightiest set of wrinkles on present-day Earth: the Himalayas.

THE DAY BECOMES FAMILIAR

We have remarked that the day has been slowly lengthening during Earth's history. The complex lifeforms that had evolved by a few hundred My ago give paleontologists more abundant ways to estimate the lengths of daily, monthly, and annual cycles than in earlier geologic periods. Just as tree-rings form once a year, various plants and animals display layers or rings of tissue formed according to daily, monthly, or yearly cycles. Their fossils are a gold mine of information as to the number of days per year and the length of other cycles.

One such study, a 1989 paper by Australian geologist G. E. Williams (see the references), indicates that by 650 My ago the length of the day and the moon's position were approaching their present values. He finds that the year contained 400 days at that time, with an uncertainty of 20 days. This means that the Earth's rotation took only 91 percent as long as it does today, or nearly 22 hours. The moon was at about 97 percent of its present distance.

Today we can actually measure the rate at which the moon is moving away from Earth and the rate at which the length of the day is slowing down. It seems that the moon is moving away by a little more than an inch each year, and the day is lengthening about 20 seconds every million years, or by a bit more than half an hour every 100 My.

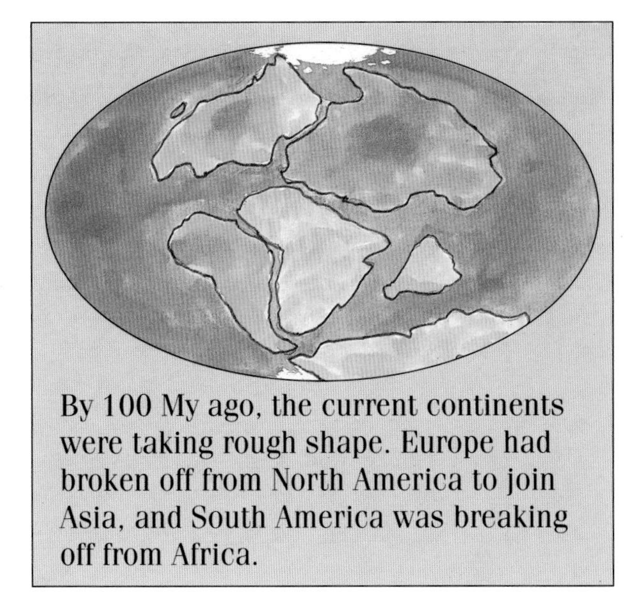

By 100 My ago, the current continents were taking rough shape. Europe had broken off from North America to join Asia, and South America was breaking off from Africa.

This rate can't be extrapolated more than 100 My or so, because the rate changes depending on the distribution of shallow oceans, which in turn affects the efficiency of forces caused by tides. Nonetheless, the measured rate suggests that 100 My ago, as the continents took their present shapes, the day was about 23½ hours long. How remarkable to realize that Earth has looked familiar and has had a more or less familiar day length for only the last 2 percent of its entire history!

Nevertheless, although Earth may have looked familiar, it was occupied by strange and ferocious beasts—the dinosaurs. And, biologically speaking, everything was about to change.

250 to 65 My Ago

12. Death and Transfiguration: The Great Land Animals

It was a weird place in itself... a rookery of Pterodactyls.... All the bottom area round the water-edge was alive with their young ones, and with hideous mothers brooding upon their leathery, yellowish eggs.... above, perched each upon its own stone, tall, grey, and withered... sat the horrible males, absolutely motionless save for the rolling of their red eyes or the occasional snap of their rat-trap beaks....

—*Sir Arthur Conan Doyle,* The Lost World, *1912*

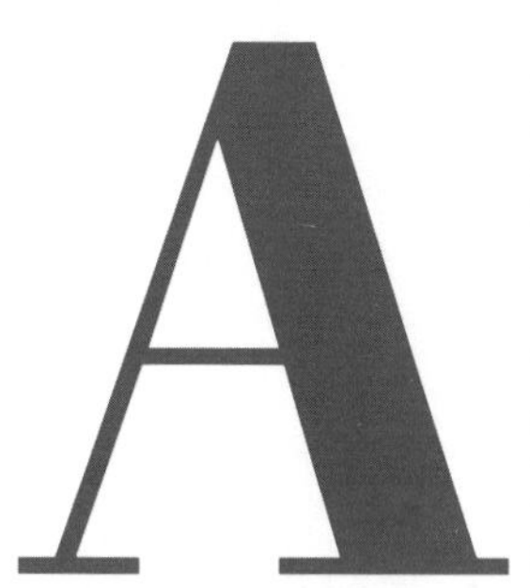

s we saw in Chapter 10, plant life colonized the shifting continents with extraordinary vigor around 460 to 400 My ago. A wave of green spread over brown landscapes in only 50 My, and animals were quick to follow. Only in terms of human time scales does the rate of proliferation of life seem slow; compared with the planet's time scale, it was rapid and recent. Only during the last 9 percent of Earth's history would an imaginary time traveler have been able to

250 My 65 My

320 My ago. Giant insects of the Carboniferous forests. The land surface was at this point widely covered by genuine forests, whose fallen logs and other plant material eventually formed many of the coal and oil deposits that powered the Industrial Age. Half of Earth's oil reserves also originated from the decay of microorganisms deposited during a 30-My interval in the midst of the Carboniferous.

250 My ago. Aftermath of the Great Dying. At the end of the Permian Period, around 90 percent of the species, including some early, large dinosaur-like reptiles, died out mysteriously. Geologists and paleontologists are still trying to determine the reason.

see abundant plant and animal life on the land. No wonder it is hard for us to follow the tangled net of evolution of various species of invertebrates and vertebrates that led to humanity: it all happened so fast.

We are still living through the period of revolutionary change.

INSECTS, VERMIN, AND AMPHIBIANS INVADE THE LAND—400 TO 200 MY AGO

Animals lagged only a little behind the plants, spreading onto land within a few tens of millions of years of the plant pioneers. They could hardly ignore a vast environmental niche covered with uncontested food.

Paleontologists have long known that the earliest land animals appeared roughly 400 My ago, but this date keeps getting fine-tuned due to new fossil finds. Based on discoveries of fossil spiders and centipedes in England, discoveries announced in 1990, the date has been shifted back to 414 My, and there is some evidence for dates as early as 430 My from fossils in Pennsylvania. A good estimate for the date when our imaginary time traveler might first have been likely to notice animals creeping amidst the shrubby early landscapes would be 410 My ago.

These first land animals were not our favorites. The earliest well-described fossils of land animals include those of spiders, millipedes, scorpions, and some wingless insect varieties. Curious, isn't it, that these first continental colonists are among the crea-

tures instinctively loathed by us? Perhaps they have been the natural enemies of all the succession of creatures who emerged later to battle the spiders and scorpions for a place on the land. The latecomers were the vertebrates, or animals with backbones—amphibians, reptiles, birds, and mammals, including human beings, who emerged only in the last few My. Despite the self-proclaimed majesty of the vertebrates, insects became the masters of the land and have remained so. About three-fourths of all living animal species are insects. Cockroaches, for example, are among the oldest living winged insects; they have remained nearly unchanged for some 320 My! Anyone who battles the ancient cockroaches in his or her kitchen learns something about insect tenacity.

Nonetheless, the most interesting evolution for us to follow is not that of the insects but that of the more obscure vertebrates.

Vertebrates descended from the fishes; in terms of skeletal structure, they are walking, air-breathing fishes. The first vertebrates to struggle onto the land were the amphibians of the late Devonian Period, about 360 My ago. From their structure, we can see their clear relation to the lobe-finned fishes, whose fins attached in a way that permitted a free swivel movement by means of muscles that extended into the fin. A famous example of the lobe-

230 My ago. A typical landscape during the Triassic Period, characterized in many areas by warm seas and dry deserts. On land the amphibians are beginning to lose ground to the reptiles, who are less dependent upon water.

finned fishes, the coelacanth, is very common in fossil records through the Cretaceous Period up to 65 My ago, but was long thought to have been extinct since then. However, in 1938, a South African fisherman caught one! Several dozen have been caught since, giving scientists a valuable window into what had seemed like the irretrievable past.

A modest evolutionary leap permitted some descendants of the lobe-finned fishes to crawl onto land: the lobe-shaped fins doubled as stubby legs. Many of these fishes lived in freshwater streams or lakes that seasonally dried up. The survival value of their crude legs was probably not so much in the search for food on land as in crawling or flopping from one shrinking pond to a larger nearby pond, where the animal might survive the dry season.

This is a good place to note that the driving force in evolution is *not* some a priori realization by a meta-brain that a certain development would work. It is not that a certain mutation, such as legs, can be seen in advance to be a good idea and therefore evolves. According to the modern point of view, neither the animal itself nor some instinctual life force of the species senses that a certain design modification would help and then ordains it. Rather, new traits arising from mutations (such as a more muscled fin) cause individuals bearing that trait to survive longer and thus to have more offspring. The mutations are not lightning-bolt creations of new species, but normal variations that are present in the population all the time. If the right environmental niche presents itself, individuals having the trait will exploit it and will leave offspring exploiting it. And enough of these offspring reproducing that trait may build up a population that becomes a new species.

Apparently the impetus of seasonally drying freshwater ponds led to the proliferation of another crucial adaptation among lobe-finned fishes, the evolution of lungs permitting the breathing of air. This is no strained reconstruction: lungfishes still live in ponds in the Southern Hemisphere and breathe air when the water becomes too fetid to

250 My ago. Some researchers believe that Earth's region of the solar system is occasionally visited by swarms of comets perturbed simultaneously from their orbits in the outer solar system by a passing star. Here, on an evening on a shore at the end of the Permian Period, the sky is ablaze with dozens of distant comets. Impacts of several comets on Earth within a few million years might have been one factor in causing sudden changes in climate, extinctions in some species, and the emergence of other species in new conditions. No proof of this theory has emerged, though proof of impacts by comets, which are chunks of dirty ice, might be hard to uncover 225 My later.

210 My ago. Dinosaurs of the late Triassic Period cavort in a geothermal area of active hot springs. Details of various dinosaurs' behavior and preferred habitats are still being studied by paleontologists.

provide sufficient oxygen. Survival of such creatures makes the transition from fish to amphibian one of the best documented in the record of life's evolution. According to dated fossils, frogs, salamanders, now-extinct alligator-like creatures, and other amphibians were exploiting the swamps and riverbanks of Earth by 200 My ago. They paved the way for reptiles.

THE "GREAT DYING" AT THE END OF THE PERMIAN—250 MY AGO

We've skipped over an incredible event, perhaps the most remarkable and mysterious event in the history of biological evolution. In the midst of some of the changes mentioned above, life was nearly wiped off the planet 250 My ago!

The evolutionary road from the earliest amphibians to the later great reptiles was not smooth. There were ups and downs, and individual species flourished and then declined. But something unusually destructive happened 250 My ago. At that time, more than half the biological families of shallow-water marine animals—or to put it another way, 90 percent of the existing species—disappeared within a few million years.

To appreciate such statistics, it helps to understand the biological classification scheme, introduced in its original form around 1750 by the great Swedish naturalist Carl von Linné (1707–1778; also known by the Latinized form of his name, Linnaeus). Linné amassed a collection of newly discovered plant and animal specimens from traders who were then opening new frontiers, and he came up with a classification scheme:

Kingdom (e.g., Plant, Animal, Fungi)
Phylum
Class
Order
Family
Genus
Species
Individual

Many individuals can belong to one species, many species to one genus, and so on. Thus, in principle, all humans but a handful could be wiped out and the species could still be preserved. Ninety-nine percent of the species in one genus could be wiped out while still preserving the genus. A sizable fraction of families in all orders could be wiped out while still preserving most or all of the orders. And so on.

This is the sort of thing that happened at the end of the Permian, when 90 percent of all species died out. This event is called the Great Dying. It marked the end of the Permian Period and the beginning of the Triassic. Trilobites disappeared. Most families of the common clam-like shelled creatures, the brachiopods, disappeared. Three of the four orders of amphibians then present disappeared. We know that it happened within a few million years, but we can't tell if it happened in a much shorter interval because it is hard to resolve short time intervals in the often-confused fossil record.

In any case the Great Dying was the most radical known biological restructuring in Earth's history. So radical was it that geologists use it not only to define the end of the Permian Period (about 290 to 250 My ago) but also to define the end of the whole Paleozoic Era, meaning the era of ancient life, about 670 to 250 My ago.

Considering what an amazing event the Great Dying was, it is surprising how little attention it got in classical geology textbooks. Many geologists of the last generation thought (or hoped) that these changes were just part of the normal (slow, uniformitarian) flourishings and failings of the various species, and they tended to play it down. A widely used 1963 text in historical geology gives only about half a page (out of 476) to this event. The twenty-six-page chapter on the Permian Period stresses mountain-building and continental drift during this period, and consequent climate change, and hints vaguely that these changes might have caused the extinctions. But the author (geologist Carl Dunbar) wants the changes to be gradual and uniformitarian; he is loath to recognize them as abrupt. The final words of the chapter: "This was a time of rapid evolution. . . . The net result was the disappearance of many of the characteristic groups of Paleozoic life, but the change was orderly and gradual, not cataclysmic."

One reason geologists of earlier generations downplayed the Great Dying was that they had little concrete evidence about it. Uncertainties in the older data allowed the possibility that the extinctions were drawn out over a long period. Less valid reasons were that it didn't fit their preconceptions about smooth geological evolution and smooth biological evolution.

By 1988, however, Preston Cloud's text on historical geology deals with the Great Dying as a rapid collapse of the ecosystem, and Cloud could write: "The collapse was awesome." The modern question is not *whether* something unusual happened but *how* it happened. Scientists have struggled to come up with an explanation for the Great Dying. Cloud lists eight possible causes: glacial cooling; a general warming; an upwelling of oxygen-poor waters, due to a change in ocean flow, that killed sea life; a reduced number of environmental niches due to shoreline changes as continental drift proceeded; the elimination of a few key species in the food chain, knocking out other species; appearance of natural chemical poisons in the environment; irradiation from a supernova; or collision of Earth with an asteroid or comet.

The fact is, no one really knows what caused the Great Dying 250 My ago.

There was a second, less significant dying 65 My ago. Being more recent, it is much better documented. In the 1980s new evidence emerged that seems to solve the mystery of what caused this more recent event. We will save that discussion for a moment, and return to the whole question of extinctions in the next chapter.

THE EMERGENCE OF THE GREAT REPTILES

After the Great Dying at the end of the Permian Period, so many species had disappeared that there

160 My ago. The great coniferous forests of the Jurassic Period. Before flowering plants evolved, coniferous forests similar to those of today flourished in many regions of Earth. Not every Jurassic landscape was crowded with dinosaurs!

250 My 65 My

150 My ago. Pteranodons, flying reptiles with wingspans of around 7 meters (23 feet), soar over a beach at the beginning of the Cretaceous Period. Pterosaurs, the general group of flying reptiles, evolved alongside the dinosaurs, but died out with them at the end of the Cretaceous Period 65 My ago.

Hartmann

were many environmental niches waiting to be exploited. Thus began, 250 My ago, a whole new era of life, the Mesozoic Era, or middle era. The whole pattern of competition had been changed by the Great Dying. Minor species that previously had had no chance to compete against dominant species suddenly found that the former "masters" had disappeared. It was in this context that the reptiles expanded so mightily.

It was a vivid transformation of life. The Mesozoic Era became famous for the reptilian monsters that emerged to take over the land: the dinosaurs. The Mesozoic consisted of three short geologic periods, beginning with the Triassic Period (about 250 to 200 My ago) and lasting through the Jurassic and the Cretaceous Periods. The end came with another, lesser dying, mentioned above, at the end of the Cretaceous Period 65 My ago.

140 My ago. A plesiosaur looms in the waters of the Jurassic seas, feeding on a school of small fish. Many plesiosaurs averaged about 4.5 meters (15 feet) long, but later types, as illustrated here, had longer necks and grew to lengths as great as 13 meters (43 feet).

250 My 65 My

Six orders of reptiles had already existed in the Permian, and all six survived the Great Dying with at least some representative species. They became the founders and ancestors of the new reptilian world. However, as a reminder of the devastation, note that there were fifty genera of mammal-like reptiles in the Permian and only one of *those* genera (called *Dicynodon*) survived. It was those mammal-like reptiles that were destined to produce *us*. "So close did we come," Preston Cloud comments, "to having our ancestry cut off at the roots."

The Mesozoic is the era that is usually represented in our images of ancient Earth, and it is the large reptiles of the Mesozoic that became the archetypal prehistoric monsters. Mesozoic reptiles were an advance over the creatures of the Paleozoic Era when it came to colonizing the waiting continents. Amphibians return to the water to reproduce because their eggs cannot withstand the drying effects of the air. This was clearly a limitation on their exploitation of the potential food resources of the land. The big advance among reptiles was that after fertilization within the female body, the eggs develop tough shells, so they can be laid in nests on the land. This meant that reptiles could wander ever farther from sources of water, spreading out to forage for the plants that had already spread across the continents, and escaping the predators that sought the eggs and young of the shore-bound amphibians.

Thus, as early as 240 My ago, reptiles marched across the land, proliferating into many diverse types. Among those types was the most famous of prehistoric beasts—the dinosaur.

DINOSAURS

Dinosaurs have been recognized for little more than a century. Various lizard-like fossil bones were recognized in Europe and America after 1770, but the name *dinosaur*, meaning "terrible lizard," was not coined until 1841 by a London naturalist, who recognized that many of the bones were far too big to belong to ordinary lizards. Dinosaurs were popularized in 1854 when life-sized models of the creatures were shown in an English exhibit opened by Queen Victoria herself. The monsters thrilled the British public. England was the perfect place to introduce dinosaurs, for the English have always been beguiled by animals real and fanciful: they gave us Winnie the Pooh, the Cheshire Cat, the Jabberwock, Toad of Toad Hall, and other such creatures. The difference is that the dinosaurs, though just as bizarre, were real.

Dinosaurs quickly stimulated a whole field of art, sometimes called paleoart. Charles R. Knight (1874–1953), working primarily at the American Museum of Natural History in New York, and other artists produced awesome images of an ancient world populated by vanished monsters standing several times taller than a man or a woman. Generations of children and grown-ups have been fascinated by these paintings ever since. "Artists are the eyes of paleontologists, and paintings are the window through which nonspecialists can see the dinosaurian world," wrote Canadian paleontologist Dale Russell in 1987.

The most eerie thing about re-creations of dinosaurs is the realization that these huge creatures were thrashing about the countryside for more than a hundred million years with no one around to see or hear them. Contrary to countless comic strips and many movies, no humans were alive during the age of dinosaurs—humans were still 200 My in the future.

Following the biological changes as the big reptiles emerged from the amphibians is tricky because their bone structure is similar; the main difference is the hard-shelled egg that permitted reptiles to roam the continents. The first egg-laying reptiles apparently emerged as early as 300 My ago, before the Great Dying. In fact, the Great Dying may have encouraged their spread by wiping out many of their competitors. Soon afterward, reptiles overwhelmed the remaining large amphibians, who had reached the size and shape of fat alligators.

100 My ago. The extent of the pterosaurs' flying ability is uncertain. Like people on hang gliders, some species may have soared to high altitudes on thermal updrafts.

Since the Triassic, amphibian life has been relegated primarily to small creatures such as salamanders and frogs, who live unobtrusively near the water.

Dinosaurs spread over the available landmasses, and truly ruled all the continents. In 1991, for example, geologists discovered the first dinosaurs known from the mainland of Antarctica. These were found only 400 miles from the present-day South Pole. The 200-My-old fossils from this site confirm not only the spread of the dinosaurs but also the drift of the continents. They indicate that 200 My ago, this region had a warmer climate with forests and flowing rivers; Antartica was at low latitudes and part of mighty Pangaea, which was breaking up as described in the last chapter.

250 My 65 My

This period, 200 My ago, was the heyday of the dinosaurs, which were proliferating into many different types. Those listed below are among the best known.

Sauropods: Long-necked herbivorous dinosaurs, the largest land animals ever to evolve. They included the enormous brachiosaurus (once known as the brontosaurus). With its graceful, arching neck and long tail, the brachiosaurus has perhaps the most familiar dinosaur profile. These dinosaurs were once thought to spend most of their time in water to support their unwieldy bulk. More recent studies of their musculature suggest that they spent much of their time on land.

Stegosaurs: Herbivorous dinosaurs with bony plates on their back.

Horned dinosaurs, or ceratopsians: Herbivorous dinosaurs with horns and backward-projecting armored shields over the neck. Triceratops is a famous example.

Bird-hipped dinosaurs, or ornithopods: Herbivorous dinosaurs with a bird-like pelvic structure. Many walked on two legs. Iguanodon is a well-known ornithopod.

Theropods: Meat-eating dinosaurs. Tyrannosaurus is the most famous example, and the most ferocious land-dwelling carnivore ever to evolve.

A generation ago, dinosaurs were usually visualized as slow-moving, cold-blooded, and stupid creatures, sluggish denizens of warm swamplands. But dinosaur behavior has always been controversial.

100 My ago. During the Jurassic Period, Earth's landscapes were beginning to take on their present form, except for the dinosaurs that were in their heydey. Birds had recently begun to evolve from soaring reptiles, and only if we saw them close up would we perceive their still-grotesque reptilian appearance.

Researchers in the late 1800s recognized similarities between the bone structures of dinosaurs and birds, and some of them visualized the dinosaurs as active and nimble. In recent decades, that view has reemerged. The bird-hipped dinosaurs in particular may have been frisky; we can picture them hopping or running after their prey like modern ostriches.

A growing, but still controversial opinion is that dinosaurs were warm-blooded and agile. Paleontologist Robert Bakker, after studying size and footprint patterns of dinosaurs and modern beasts, concluded that some dinosaurs could move at speeds comparable to many modern predators. In a 1987 article, he noted:

> Marks left by ligaments and muscles show that dinosaurs had great power at shoulder, elbow, hip, and knee. . . .Pound for pound, most giant dinosaurs were stronger, faster, and more maneuverable than the rhinos and elephants of today.

Not everyone accepts nimble dinosaurs. Perhaps future fossil finds will clarify the "personalities" of Earth's ancient reptilian masters.

WHAT THE BIRDS TELL US

Another dream of fossil hunters is to clarify the ancestry of birds, which are, after insects, among the most successful land colonists, with nearly 9000 known species. As can be imagined from the above discussion, in terms of their anatomical structure and their practice of laying eggs, birds are more closely related to reptiles than to other animal groups. On this basis, birds have been characterized as "feathered reptiles."

However, there is one big difference between modern reptiles and birds: reptiles are cold-blooded, and birds are warm-blooded, which means that they use some of their energy to maintain their bodies at a constant temperature.

As early as the 1800s some researchers suggested that birds were specifically descended from di-

90 My ago. The flowering plants became the dominant land-plant species fairly rapidly during the Cretaceous Period. Prior to that time, true flowers had never graced terrestrial landscapes. Today 250,000 species, or 96 percent of the vascular land plants, are of the flowering variety. The rest are ferns and other species that are remnants of the nonflowering plants that covered the land 200 to 400 My ago.

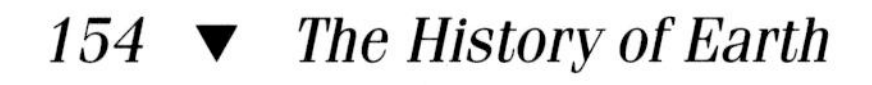

Hartmann

80 My ago. The mighty Tyrannosaurus rex, fabled as the most fearsome of the meat-eating dinosaurs, frightens away one of the lesser, bird-like dinosaur species in the midst of a dense Cretaceous forest.

nosaurs, such as the bird-hipped ones. Many contemporary paleontologists respond to the question of warm-bloodedness by asserting that at least the bird-hipped dinosaurs were warm-blooded. The same researchers believe that dinosaurs' nesting behavior was similar to that of birds—in keeping with the new picture of a more lively dinosaur lifestyle, and in keeping with a direct link to birds. A growing consensus is that modern birds are evolutionary offshoots of dinosaurs or dinosaur-like species.

One of the big problems in tracing birds' ancestry is that bird bones are thin and delicate, in order to allow flight, and therefore make poor fossils. Hence, bird fossils are rare and it is hard to follow the evolution of birds through an unbroken chain of fossil evidence. It is ironic to think that our friends, the innocent songbirds, may be the descendants of the monsters of the Jurassic.

WAXINGS AND WANINGS

The era of dinosaurs was not static, with dinosaurs springing into existence at the beginning of the Triassic Period and remaining as a constant group of species throughout the Cretaceous. Like other species, they had their successes and failures. The stegosaurs declined in the early Cretaceous, perhaps 140 My ago, while the horned dinosaurs appeared fairly late, perhaps 130 My ago, in the mid-Cretaceous.

All the while, the mousy early mammals lurked in the background, probably fearing the mighty tread of the heavy reptiles. If you put birdseed in your backyard feeder, there is a lot of history encapsulated in the contest that you may witness between the birds and the squirrels—a last vestige of the age-old battle between the dinosaurs and the mammals over the food supply of the land, a battle about which we will say more presently.

Except for the stegosaurs, most dinosaur types made it through to the end of the Cretaceous, 65 My ago. Some of them may have been on the decline by that time. But at that moment in geologic time, something disastrous happened. It ended not only the Cretaceous and the dinosaurs, but also the whole "middle life" biological framework of the Mesozoic Era. That disaster is the subject of the next chapter.

65 My Ago

13. The Amazing End of the Dinosaurs

A minor accident forced [my plane] down in the Rio de Oro region, in Spanish Africa. . . . I found myself on . . . a plateau. . . . The surface sand [was] composed of minute and distinct shells; . . . The first star began to shine. But suddenly my musings on this white sheet and these shining stars were endowed with a singular significance. I had kicked against a hard, black stone. . . . A sheet spread beneath an apple-tree can receive only apples; a sheet spread beneath the stars can receive only star-dust. Never had a stone fallen from the skies made known its origin so unmistakably.

—*Antoine de St.-Exupéry,* Wind, Sand, and Stars, *1939*

During the 1980s a tremendous debate raged over the fate of the dinosaurs. Long before then, geologists had known that the dinosaurs and many other Cretaceous species died out fairly abruptly, only to be replaced by mammals and other more familiar plants and animals of our era. Some 75 percent of known species died out within a few million years! The extinctions were especially severe among sea creatures. As with the Great Dying at the end of the Permian Period, nobody talked much about why this dramatic change occurred because there was not much evidence about it. Most paleontologists had tacitly assumed that this was yet another example of evolution by natural selection: through competition for food, the small, early mammals had exploited an

70 My ago. In the forests of the late Cretaceous Period.

environmental niche that favored agile, intelligent creatures. They had outmaneuvered the big, supposedly sluggish dinosaurs. In taking over the world, they had confirmed Darwin's theory that evolution is driven by competition among species. One problem with this view was the emerging revisionist idea that dinosaurs were more agile and better adapted to their world than had been thought.

In 1979 and 1980, a heretical new theory was proposed. A father-and-son geophysics team made up of Luis and Walter Alvarez, together with their colleagues Frank Asaro and Helen Michel from the University of California at Berkeley, published evidence that an asteroid hit Earth 65 million years ago, causing a titanic explosion and the ensuing climatic changes that wiped out the dinosaurs and many other species. In short, the transition between the Cretaceous and the Tertiary periods had come because of a full-blown catastrophe.

What a shocking idea this was to traditional geologists! Remember that modern geology was born only by overturning ideas of sudden catastrophic changes in Earth's history. Lyell, the founder of geology, in his famous 1835 textbook, had lampooned exactly such ideas:

> We hear of. . . the sudden annihilation of whole races of animals and plants, and other hypotheses, in which we see the ancient spirit of speculation revived. . . . In our attempt to unravel these difficult questions, we shall adopt a different course, restricting ourselves to the known or possible. . . .

For a century, most geologists had restricted themselves to Lyell's uniformitarian ideas: slow and steady geologic changes due to familiar terrestrial forces were "known or possible"; sudden catastrophes were anathema. Yet according to the new data of the Alvarez team, a catastrophe originating *outside the planet* was the real cause of the dinosaurs' extinction.

WHAT IRIDIUM TELLS US

What was the evidence for this radical neo-catastrophist idea? As with many great discoveries, this one was an accidental by-product of other research. The Alvarezes had been studying the abundance of the element iridium in sediments, with the relatively modest goal of measuring the accumulation rate of these sediments. During this work, they stumbled on whopping concentrations of iridium in the inches-thick 65-My-old soil layer found at the boundary between sediments formed at the end of the Cretaceous Period and younger sediments deposited on top of them at the beginning of the Tertiary Period. This boundary between the Cretaceous and Tertiary layers is called the K-T boundary, because the Cretaceous is abbreviated K in geologic mapping parlance and the Tertiary is abbreviated T. In the so-called K-T boundary layer, iridium was 20 to 160 times as abundant as in other sediments. The Alvarez team's initial sample came

66 My ago. Prelude to disaster. An asteroid destined to collide with Earth makes a pass close to the Earth-moon system (in the distance, upper center and right). Due to nearly intersecting orbits, the intruder may have made many passes near Earth before finally hitting our planet.

65 My

65 My ago. The star of doom approaches. In the last hours of the Cretaceous Period, the approaching asteroid would have been a bright object in the sky on one side of Earth.

from a well-defined clay layer at the K-T boundary in Italy, but soon various researchers confirmed the high iridium concentrations in the K-T boundary layer all around the world.

The researchers quickly realized that some worldwide source of iridium-enriched dust must have been created just during the Cretaceous-Tertiary transition 65 My ago. What could be the source of such material? Soon they recognized that the most common types of meteorites are enriched in iridium. If a large enough meteorite of such a type hit, it would explode and eject an enormous spray of iridium-enriched dust, which would be lofted into space, fall back into the upper atmosphere over much of Earth, and then be spread by high-altitude winds, not unlike the dust from the (much smaller) nuclear test explosions of the fifties and sixties. The dust would then filter down from the high atmosphere, forming an iridium-enriched sediment layer.

As early as their first full paper on the subject, published in June 1980, the Alvarez team estimated the total amount of iridium in the worldwide K-T boundary layer and calculated how big a meteorite must have hit to provide that amount of iridium, about 200,000 tons. The answer was that the meteorite would have to have been about 10 kilometers (6 miles) across! This figure has been supported by subsequent studies.

A 6-mile-wide meteorite! Those readers who have examined fist-sized stones from space in planetarium display cases may find a city-sized meteorite preposterous. But we know they are out there in space, lurking. Thousands of asteroids and comet nuclei of this size are known, and dozens of them are on orbits that approach close to Earth. In fact, astronomical calculations based on numbers of known asteroids suggest that Earth *should* be hit by one this size every hundred million years or so. So the idea, while heretical to geologists, was not so fantastic to astronomers.

WHAT THE 65-MY-OLD SOIL LAYER TELLS US

Nonetheless, many paleontologists and volcanologists did not accept the new theory at first, arguing that the iridium excess could have come from some other cause, such as an unusually massive volcanic outburst. But further studies of the K-T boundary layer turned up much more evidence. Shock-altered quartz fragments—fragments whose microscopic structures are altered in a way that happens only when quartz is subjected to sudden high pressure in an enormous explosion—were found in the layer. The shocked quartz is found in debris of all major meteorite impact craters but not in volcanic debris. Also common in the layer were tiny glassy spherules, a type of material caused as molten droplets of melted rock shoot out from an explosion and then resolidify into glassy beads. Both these materials testify that the K-T boundary layer consisted partly of debris from a mighty explosion, consistent with an asteroid impact.

Hartmann

65 My ago. Two seconds before the end of the Cretaceous. A 10-kilometer asteroid streaks through the atmosphere in about 10 seconds and is about to strike the surface.

Hartmann

65 My ago. One minute before the end of the Cretaceous Period. The forests and jungles of the world below are teeming with reptilian life.

The soil in the K-T boundary layer harbored still more secrets. In some regions it contained chaotic, tumbled rocks—telltale deposits left by a tsunami (popularly but incorrectly known as a tidal wave). In other words, a giant wave had swept over many regions, leading many theorists to speculate that the impact had happened in the ocean.

Even stranger, by the mid-eighties, scientists discovered that the K-T boundary layer also contained soot. The soot component was found all over the world. Researchers calculated that to generate the amount of carbon in this soot layer, you would have to burn down more forest than exists in the whole world today! Of course, the Cretaceous world had more forests than we do, for much of our timber has already been cut. Scientists studied the amount of soot produced per unit of plant material burned and found that this amount depends strongly on the conditions of burning; however, they concluded that most of the world's forests burned at the end of the Cretaceous! Again, this is strong evidence of an event much different from any plausible volcanic eruptions.

Despite many scientific arguments back and forth, the idea of a catastrophic impact 65 My ago now seems well established. As early as 1983, *Science* reported that "for many, at least, asteroid impact has been accepted as a causative agent in mass extinction." And by 1990, geologist V. L. Sharpton of the Lunar and Planetary Institute in Houston was able to remark that while workers in other fields may still be arguing, geologists who are familiar with all the features of the K-T boundary layer now accept asteroid impact as the main cause of the K-T event with almost "divine certitude."

THE EXPLOSION—AN EYEWITNESS ACCOUNT

I described the event that took place 65 My ago as a "titanic explosion" because it was the most violent event in the last 100 My and would wipe out civilization if it happened today. However, we must remember that it was a trivial event when compared with earlier cratering on Earth and the moon. Remember that in the granddaddy of all impacts—the giant impact that formed the moon 4500 My ago—Earth was hit by a Mars-sized impactor, perhaps 6000 kilometers across! Many lunar basins, giant impact scars still visible on the long-dead moon, involved impacts of asteroids 150 kilometers across. Indeed, in the earliest days of Earth's history, so many asteroid-like bodies were falling on the planet that puny 10-kilometer bodies—the kind that wiped out the dinosaurs—hit yearly or perhaps even monthly!

Nonetheless, the impact at the end of the Cretaceous, only 65 My ago, was the most dramatic event in the recent history of Earth, and its discovery has been one of the most exciting chapters in science in the last decade. It is worth reconstructing an "eyewitness account" in some detail, not only because it was a dramatic event of the past but also because it could happen again in the future, as we will see. We can get a fair idea of the event by applying data on asteroids' properties as well as knowledge of impact explosions drawn from H-bomb tests and computer models.

Imagine that we are floating about 100 kilometers above the surface on that fateful day 65 million years ago. Our vehicle of the imagination is a cross between the space shuttle and a time machine. For days, we have been seeing a slowly brightening "star" in the morning or evening sky. The asteroid is approaching on a collision course but is still too far away to have a visible shape.

Below us, continental coastline details are more or less unfamiliar. Only in the coming millions of years will erosion and plate tectonic disruptions alter them to the exact profiles seen in today's maps. From this altitude, we get no hint that the land surface below teems with reptilian life. Fearsome di-

Hartmann

65 My ago. One minute after the end of the Cretaceous. Debris from the impact is hurtling into space from the impact crater.

65 My ago. In the first minutes of the Tertiary Period, the distant explosion caused by the asteroid impact might have seemed only a momentary disturbance to the flying reptiles of the day.

nosaurs roam. Scenes of reptilian life and death are played out in the blue-green haze far below. Great insects buzz among fantastic plants.

Only within the final hour before impact has the asteroid drawn close enough for us to look upward and make out its irregular round shape. In the final fifteen minutes it moves across the sky, then rushes past us and enters the upper atmosphere.

A fascinating thing about huge impacts is their slow time scale. The impact of a bullet or a rocket is over in a flash, literally. A camera running at several thousand frames per second is needed to record such events in slow motion so that we can see them. Even the rapidly rising fireball of an H-bomb explosion, filmed at ordinary speed, can only hint at the terrible, slow majesty of the impact that ended the reign of dinosaurs.

A typical asteroid arrives at about 15 to 18 kilometers per second. At such a speed, the asteroid takes eight to fifteen seconds just to traverse the atmosphere, depending on the angle of passage. Friction with the upper atmosphere heats it, blows off loose dust, and eventually vaporizes the leading edges at altitudes around 150 kilometers. A brilliant fireball forms around the leading side, and a strange contrail extends behind, as the object takes on the shape of a fiery comet.

Perhaps the object is as strong as many solid-rock meteorites, and it reaches the surface intact. But it may be one of the types of asteroids descended from a comet nucleus—weak, crumbly, perhaps containing some ice. Such crumbly meteorites often break up in flight, producing a shower of many fragments that fall at different locations, especially if the flight path is at a shallow angle to the ground. But the size of the 10-kilometer asteroid is an appreciable fraction of the thickness of the main part of the atmosphere, and so it is unlikely that it would have had time to separate into fragments hitting at different spots unless it had come in at an extremely horizontal angle.

The shaft of the 10-kilometer-wide contrail can expand at no more than the speed of sound. Thus during the eight-second flight, its top part grows no more than a few kilometers wider than the base at the impact site. It forms a striking cylindrical column through the atmosphere, hanging for a few moments like the finger of an angry God reaching through the whole atmosphere to touch the planet with death.

In the final second of the Cretaceous Era, the asteroid penetrates the surface. Perhaps it hits an ocean surface, but its diameter is several times thicker than the 4-kilometer mean depth of the oceans, so that any water it encounters is in effect merely a thin surface film. It penetrates the ground (land or sea floor) with a kinetic energy equivalent to a hundred million one-megaton H-bombs.

Within a minute the asteroid has disappeared, transformed into an exploding fireball. Under the ground, the asteroid has disintegrated in a mass of exploding steam and luminous, vaporizing soil, which expands outward and upward during the next minutes, excavating a crater that will eventually approach 200 kilometers in diameter. A conical sheet of pulverized debris shoots upward from the rim of the expanding crater into the atmosphere. There may also be a higher-speed vertical plume of spray, as observed in high-energy impact experiments.

Also in the first minute, the shock wave of pressurized air, spreading through the atmosphere at the speed of sound, has spread out to about 18 kilometers. If the impact happened in the ocean, the shock pulse traveling at the speed of sound through ocean water has reached a distance of no more than about 65 kilometers. Giant tsunami waves in the ocean lag behind this, because of their slow velocity. However, the seismic waves traveling in Earth's solid crust have reached perhaps 320 kilometers, creating earthquakes as they go. Animals at this distance on nearby continents still do not know what has happened, but they have felt the tremors and might be raising their heads to see the distant pale curtain of debris rising from beyond the horizon, suddenly screening the sun and darkening the sky—a shadowy harbinger of more destruction to come.

WHERE DID THE IMPACT OCCUR?

Within a day, the K-T impact created a crater probably close to 200 kilometers across. Shouldn't a crater this size be easy for modern geologists to find? If it was well preserved on a stable landmass or shoreline, it would be one of the most dramatic features of our globe. Doesn't its absence prove that the K-T impact theory is wrong?

Not necessarily. The key word here is "stable." Remember that Earth has extremely active geology. Rains erode the continents and the restless crustal shiftings of plate tectonics crumple the edges of colliding continents and drag large portions of some plates under the edges of other plates, in a process called subduction. Remember, too, that much of the Atlantic Ocean floor has been created since 65 million years ago as Europe/Africa moved away from the Americas. At the same time, large regions elsewhere on the globe have been subducted, plunging sluggishly into Earth's mantle. All their geographic features have been lost forever. Thus there is a good chance that the crater has been destroyed by these processes.

Nonetheless, there is also a chance that the site

of the K-T impact was not in one of the regions that has been destroyed. One of the detective stories associated with the K-T boundary is the search for the site of the impact. Could the crater be buried under ancient sediments or hidden on the sea floor beneath layers of sludge? As mentioned, early studies, around 1980 to 1984, suggested that the impact happened in the ocean. There are several reasons. First, oceans are the most likely targets statistically; they cover 71 percent of our planet today, and probably covered a similar fraction at that time. Second, there is no obvious crater on land, but an ocean impact would have made a sea-floor crater hidden from our latter-day prying eyes or perhaps already destroyed by subduction or masked by the deposition of sediments. Third, certain geochemical studies of the spherules and other ejected debris are consistent with a hit on a shallow sea floor. Fourth, the tsunami deposits suggest a giant ocean-spawned wave that swept over coastal areas. Finally, the especially strong extinctions among sea species also suggest that the asteroid might have hit the sea.

On the other hand, in 1984 geologist Bruce Boher and his co-workers concluded that the shocked minerals found in the K-T layer come from crystalline continental rocks, not the unconsolidated sediments typical of sea floors. In support of this view, various workers have found that the feldspar minerals found there are the low-calcium types typical of continental granites, not the high-calcium types typical of sea-floor rocks. These results argue against an impact on a deep-sea floor, but could allow an oceanic impact on a continental shelf or in a shallow offshore sea, as well as on dry land.

Taken together, the evidence for a mixture of sea-floor and continental debris suggests that either (1) the asteroid hit near a continental margin where it could penetrate not only sea-floor rock but also sediments washed offshore from continents, or (2) more than one crater formed, with both oceanic and land sites represented.

K-T sleuths have recently found more evidence

65 My

65 My ago. The impact of the asteroid created a tidal wave in the surrounding ocean that was probably many hundreds of feet high, swamping low-lying coastal areas hundreds of miles from the site of the crater.

for an offshore site, and have possibly located the elusive impact site. The evidence comes from the geographical distribution of so-called turbidite deposits, the chaotic deposits caused by the tsunami waves. University of Arizona geologist Alan Hildebrand found that these are thickest in the Caribbean and thinner as one travels north onto the southern coast of North America. This suggests that the impact happened offshore in the Caribbean.

Hildebrand mapped the thickness of the deposits of debris ejected from the crater, indicated by the shocked quartz and glassy spherules. In Haiti, he found the thickest known K-T boundary deposit of ejecta debris—50 centimeters thick, compared with more common thicknesses of only 2 centimeters or less at more distant sites. From his data, Hildebrand concluded that the impact occurred somewhere in the Caribbean.

Hildebrand then began searching for the crater itself, and by 1990 came up with a very strong candidate on the north coast of Yucatán. It is partly on land and partly under water. No crater can be seen there now because the whole structure, named Chicxulub, is buried under sediments. However, it shows up in gravity and magnetic field surveys as a feature about 180 to 200 kilometers across. Moreover, Hildebrand and his colleagues identified heavily fractured rock and shocked-quartz fragments at the site—a strong sign of a large impact. Stratigraphically, Chicxulub is near the K-T boundary, but it hasn't been proven to coincide exactly with the K-T boundary.*

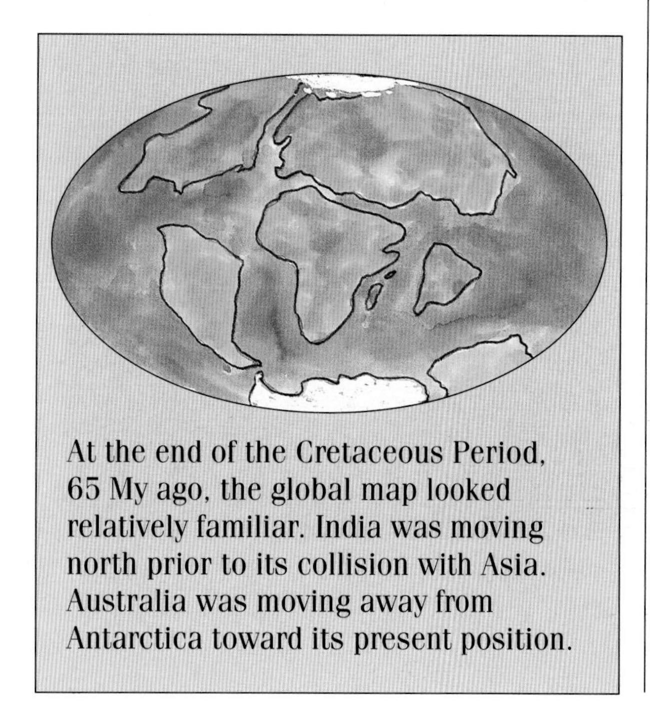

At the end of the Cretaceous Period, 65 My ago, the global map looked relatively familiar. India was moving north prior to its collision with Asia. Australia was moving away from Antarctica toward its present position.

The Yucatán site looks like the best bet, but Hildebrand and other "crater detectives" are continuing the studies in hopes of getting definitive information on the events that changed Earth's life forever.

WAS THERE AN EPISODE OF SEVERAL IMPACTS?

The original paper of the Alvarez team in 1980 assumed that only one random asteroid hit Earth and many of their followers have assumed the same thing, but scientists continue to wonder about the possibility of multiple impacts in a short interval of geologic time. This might seem statistically improbable. How could it have happened?

There are several ways. For example, what if the body that hit Earth was not an asteroid but a comet? Comets have been seen to break into several pieces spontaneously as they pass around the sun in the inner solar system. Perhaps such a comet's orbit crossed that of Earth 65 My ago, and soon after the comet fragmented, Earth hit several of its pieces within a geologically short interval of time—a million years or so. A second scenario for multiple impacts would be the chance that the asteroid initially grazed the atmosphere in a tangential approach, breaking into several pieces due to atmospheric drag. Some pieces might have hit the ground immediately, but some or all pieces might have skipped back into space, as a well-observed bus-sized meteorite did over Wyoming in 1972 (see

*For further information, see review articles by Hildebrand and Boynton (1991) and Beatty (1991, listed in the references.

65 My ago. Within an hour of the major K-T impact, swarms of debris that had been ejected into space crashed back into Earth's atmosphere, lighting up the sky, creating minor craters, and igniting worldwide forest fires.

the reference by Jacchia, 1974). If this happened, several large fragments could have gone into orbit around Earth, only to decay and hit Earth within a year or so; or these fragments could have gone into orbit around the sun, passing close to Earth, only to collide with Earth over a period of some millions of years.

Still another scenario for multiple impacts comes from observations suggesting that a few asteroids may be binary objects (one asteroid orbiting around another) or compound objects (two masses weakly connected or barely touching). If this is so, two impact craters might have been created. Such double impacts are known. The 290-My-old Clearwater Lake impact site in Canada is marked by two perfect craters, 32 kilometers and 22 kilometers across, lying side by side.

Some researchers have claimed direct evidence for multiple impacts at the end of the Cretaceous. The best indication of a second impact, in addition to the main impact in Yucatán, comes from an eroded impact crater buried beneath the sedimentary plains near Manson, Iowa. It is called the Manson structure and corresponds to a crater originally 32 to 35 kilometers across. It has been dated at 65.7 My ago—very close to the exact age of the K-T boundary—but its debris layer has not been confirmed to lie exactly on the stratigraphic K-T boundary. At least one large, eroded impact structure in the Soviet Union has also been claimed as a K-T boundary crater, but while one date for it places it at about 66 My, several other dates place it near 39 My, so that its age remains controversial. Hildebrand and co-workers also found a second Caribbean structure, resembling a crater, off the Colombian coast. It requires further study.

Some workers tout at least Manson as proof of a worldwide spate of impacts not more than a few years apart. Critics reply that Manson may have been a coincidence, a million or so years off the K-T boundary, and is too small in any case to have had global biological effect; that the Soviet crater is far too poorly dated to be included in the list for the time being; and that the Colombian feature's origin is uncertain.

There is another way to shed light on whether episodes of multiple impacts occurred 65 My ago and/or at other times during Earth's history. We could go back to the moon, which would have experienced an episode of multiple impacts that hit

Earth. Lunar craters are preserved there in abundance, and dating a large sample of them would show whether an unusual shower of impactors encountered the Earth-moon system 65 million years ago or at any other time. This would ultimately give better statistics than studies on our heavily eroded planet.

CONSEQUENCES OF THE CATASTROPHE

So far we have two basic facts about the end of the Cretaceous: a big impact occurred, and the dinosaurs and other species died out. But how did one event lead to the other?

This problem has vexed geophysicists and paleontologists for the last decade. At first some paleontologists were very opposed to the idea that the impact caused the extinctions; some argued that, from the fossil evidence, dinosaurs were already on the decline and would have died out anyway. The impact, in their view, was just incidental. But the sudden extinctions (during a period of a million years or so) of so many species now seem so dramatic that most scientists agree they were caused by the impact.

Early theories tried to explain the extinctions by a single effect of the impact, such as the blocking out of sunlight by the dust thrown into the atmosphere. Today we realize that the effects of any catastrophe, even on a regional level, are subtle and complicated. The effects of such a large catastrophe

65 My ago. A day after the major K-T impact, fires raged in many areas. Regions under heavy storm clouds may have been shielded from the celestial blast of heat. Here, on the other side of the world from the impact site, forest fires burn on the horizon, as seen from a storm-protected pocket where some species may survive.

65 My

65 My ago. One month after the end of the Cretaceous. The atmosphere is completely clouded by dust from the impact and smoke from nearly global forest fires.

Hartmann

would last over a period of time and would affect different species and different regions differently.

One group of scientists studied the effects of the tsunami wave. If the impact happened in an ocean, they calculated, it would set up a wave as much as 5 kilometers high near the impact site. As this wave spread across the oceans, its height would decline. In any case, coastal areas adjacent to the target ocean may have been inundated by a wall of water thousands of feet high. It was not a nice day to be munching leaves in your favorite coastal swamp!

The most gruesome effect was proposed in recent work by University of Arizona researcher Jay Melosh, an expert on cratering. Using computer models of cratering, he calculated that much debris was thrown out of the atmosphere into space, only to fall back again into the atmosphere. This created a rain of meteors streaking through the atmosphere with incredible concentration for a few hours. Melosh calculated the heating effect from these luminous bodies. According to his results, the sky was ablaze with meteors and the radiant energy level reached on the ground was like that inside an oven. In other words, many aboveground animals must have been broiled to death where they stood. Melosh suggested that this happened all over the world, although others think it might have been effective primarily in more localized regions near the impact. This result may explain why forests apparently burst into flame in many parts of the world.

At the same time, Melosh's blazing-sky theory reveals some of the complexities that are possible. In areas of thin cloud, the clouds would have been burned away by the heat, like fog in the morning sun. But in localized areas of thick cloud, heavy snowstorms, or driving rains, the surface may have been protected from the sudden burst of heat from above. Thus, each continent may have had pockets of animals and plants not immediately affected, areas that were spared the forest and grass fires. A week or so after the impact, these animals and plants may have begun to recolonize the surrounding burned-out landscape.

Spectacular as the coastal devastation and fires may have been, it was not the most important long-term effect, biologically speaking. Surviving plants and animals now encountered two longer-term problems.

First, according to chemical calculations, the shock of energy from the impact acted to convert

65 My ago. The largest known crater from the K-T impact is a structure estimated to have been 180 kilometers wide, ripped open at a site on the present-day coast of Yucatán. After the impact it probably refilled with water, lava, and sediments. In this view from one section of the rim, the floor is aglow with erupting lavas, while seawater begins to flow here and there over low spots in the rim.

large amounts of atmospheric nitrogen into nitrogen oxides, which promote acid rain. Thus, in addition to everything else, survivors had to contend with the sorts of pollution that damage forests in industrial North America and Europe—but at higher concentrations. Many biologists think acid rain may have been a major factor, perhaps *the* major factor, in the extinctions.

As an example of the detailed scientific work that goes into such theories, several different studies have built the case for acid rain. Geochemists suggested the nitrogen oxide production by impact shock as early as 1982. More detailed calculations on the production rate of the nitrogen oxides were done in 1987 and convinced many of the scientists of the reality of the effect. The resulting acid rain would increase the rate of erosion of continents. One result of that would be an increased deposition rate of continental sediments on ocean floors. The continental sediments are rich in the isotope strontium 87, and an increase in their deposition would cause an increased strontium 87 content in seawater. In 1988, oceanographic chemist J. D. Macdougall found that ocean drill cores do indicate such an increase at the K-T boundary, supporting the acid rain theory. Through such diverse bits and pieces of painstaking work, evidence about the K-T event mounts up.

The second long-term effect may have been more spectacular visually. According to calculations by teams at Los Alamos and elsewhere, large amounts of dust were ejected into the atmosphere. This included the high-altitude dust pall that fell back into the atmosphere from above within hours, denser clouds carried away from the impact site by winds, and a global pall of smoke from the fires. Within a week, Earth was densely clouded by dust and smoke. Day turned to night.

Similar but much thinner atmospheric dust layers have been observed from volcanic eruptions, such as the one at Mount St. Helens, in recent decades. Using careful measurements of those clouds, NASA scientist Owen B. Toon and his colleagues made computer models of the history and density of the dust pall from the K-T impact. Fed from above, the dust pall first settled in the stratosphere at altitudes of 30 to 50 kilometers, and then settled more slowly into the lower atmosphere, eventually collecting on the surface in the K-T boundary soil layer. On the basis of the fallout measured in the layer, the scientists estimated that about two pounds of dust was initially suspended above every square foot of Earth's surface. Its half-life was a few months—that is to say, about half of the dust at any one time would fall out by a few months later.

The dust pall's most astonishing effect was to block sunlight. Toon and his colleagues found that total darkness had descended. For one or two months, it was too dark to see; for two or three months, it was as dark as full moonlight; and for as much as a year, it was too dark for most plants to perform photosynthesis! The period of darkness would have destroyed the light-dependent plankton that float in the surface layers of the ocean. Destruction of this source of the oceanic food chain would have had a rapid, drastic effect on marine life in general. All this agrees perfectly with the fossil record: most life in the shallow seas was wiped out abruptly during the Cretaceous-Tertiary transition.

On land, the effects were more dragged out. Plant life as well as animal life might have survived the day in stormy regions, but the months of darkness would have been devastating. Furthermore, according to the calculations of Toon and his associates, the dust blocked enough sunlight to reduce land surface temperatures by as much as 70°F to subfreezing levels for six months or more in some areas. (This effect did not occur at sea because of the oceans' great capacity to store heat.) This effect might have been offset somewhat by a countereffect: the extra water vapor blasted into the atmosphere would have aided the so-called greenhouse effect that warms the atmosphere.

In any case, the acid rain, the darkening, and the temperature change dramatically attacked the remaining pockets of plant life. Nonetheless, many seeds and spores might have survived to regenerate life as the sky lightened after many months.

THE EXTINCTIONS

The upshot of these morbid theories is that a drastic climate fluctuation was started by the impact and lasted for some years. It would have lasted longer in some latitudes or environmental niches than in others. Various species of animals in some parts of the world and in some segments of the food chain might have lasted longer than others.

Again, this seems to fit the fossil record. There have been many arguments about whether or not some species of dinosaurs survived for an extended time after the impact. Dinosaur bones have been reported in sedimentary layers above the K-T boundary layer, leading some paleontologists to claim that some of the dinosaurs survived for a while. Others believe that these rare deposits were created by special conditions under which erosion took place. For example, consider a valley between two hills. The dinosaurs might all have died during the K-T impact, leaving bodies in the valley and on the hills. Their bones might be preserved in sediments. But during the next thousand years, bones from the hills might have been washed down along with the sediments, and have been deposited *on top* of the K-T boundary layer in the valley. Geologists

65 My later, looking at this site, would find dinosaur bones in strata above the K-T layer.

In any case, a 1982 summary of the fossil evidence by Canadian paleontologist Dale Russell concluded that the K-T "biological crisis . . . caused about 75 percent of the previously existing plant and animal species to disappear." He found that no species of land animal weighing more than about 55 pounds survived into the Tertiary, and that plants of the northern temperate latitudes were more damaged than those of the south, perhaps consistent with a Caribbean impact site.

Determining the specific sequence of extinctions and their durations during the few million years around the time of the impact is complicated by a background of random emergences and extinctions of species. Throughout geologic history, some species are in the ascendant and others on the decline, due to more ordinary causes such as climatic shifts caused by continental drift and ice ages. An exciting task awaits the next generation of geologists and paleontologists as they unravel the exact, grim history of extinctions caused by the impact itself.

ASTEROID IMPACTS: IMPLICATIONS FOR BIOLOGICAL EVOLUTION

The attention to the K-T impact caused more analysis of the effects of asteroid and comet impacts in general and of the nature of biological change. The famous biologist Stephen Jay Gould remarked that the impact theory of extinctions is a radical revision of Darwinian biology, because Darwin argued that evolution was propelled only by competition among species and the resulting natural selection. In Darwin's original view, evolution was driven only by forces internal to Earth's ecosystem itself. If we now admit that chance astronomical influences from outside Earth can wipe out 75 percent of the species, wiping the competitive slate clean and allowing previously minor species to gain a foothold, then this is a dramatically new view of life.

The history of life on our planet (and other planets?) might be viewed not as a long, slow, inevitable march toward intelligence, but as a series of accidents that created new biological opportunities at random. Perhaps, to put a slightly different slant on it, intelligence was able to evolve only because cosmic accidents occasionally cleared the way for new species, in the same way that occasional forest fires clean out the old underbrush and dead timber, making space for a vigorous fresh start.

In other words, the discoveries about the K-T boundary have produced not just another new scientific theory but a new philosophic view of the role of life and intelligence.

Clearly, an important question is, How often did this sort of thing occur? Could impacts really have been a major influence on biological evolution throughout the last half of Earth's history?

To answer these questions, we need to think a little more about how big an impact we need. As we said, the total amount of iridium in the K-T boundary layer indicates an asteroid perhaps 10 kilometers across, or an even bigger comet up to 20 kilometers across. The object might have been even bigger if some of the iridium was blown away into space, instead of falling back on Earth. The impact velocity is uncertain. From these figures alone, we can calculate that the crater might have been as large as 200 kilometers across. However, at least two craters originally close to 100 kilometers across, formed during the 550 My since the Cambrian, have been found: the 210-My-old Manicouagan ring in Canada, and the badly eroded Popigai feature in the U.S.S.R., whose date is uncertain. They caused lesser, if any, extinctions in the biological record (see discussions later in this chapter). Therefore, the K-T boundary crater may have been even bigger. Researcher Alan Hildebrand and some other searchers for the K-T crater favor a diameter closer to the 200-kilometer figure. So our question can be rephrased: How often do 100- to 200-kilometer craters form on Earth?

We have three independent sources of information on the rate of asteroid/comet impact and the production of large impact craters. Lunar crater counts, together with ages of lunar rocks brought back by astronauts and Soviet unmanned probes, allow us to calculate the rate of impacts on the moon in the last 3000 My. Counts of large, eroded craters on ancient terrestrial surfaces give us rates of large impacts on Earth in the last few hundred My. Finally, statistics of asteroids on Earth-approaching orbits allow us to calculate the rate of impacts in the Earth-moon system generally.

These three data sources give roughly consistent figures and imply that the cratering rate has been roughly constant in the last 3000 My. Canadian cratering expert Richard Grieve gave one of the most authoritative analyses in 1990. He tabulated known craters and estimated that craters larger than 20 kilometers form about every 3.6 My on Earth on the average. Taking into account the fact that 100-kilometer craters are only about one-twentieth as common as 20-kilometer craters, we'd expect a 100-kilometer crater every 73 My and a 200-kilometer crater every 268 My. These crude figures are consistent with the fact that the last event of K-T magnitude happened 65 My ago.

The bottom line is that Earth may plausibly have experienced one to five such events during the last 500 My of post-Cambrian time, the period from which we have good fossil records.

Why haven't we discovered more such events in the stratigraphic record? One possible answer is that there really was only one impact this big. Another possible answer is that there were several, but the older ones are harder to detect: the older the crater, the more degraded the geologic evidence. Plate tectonics may have completely destroyed the older craters; they could have been subducted into the mantle. And their iridium-rich ejecta or other debris may have been more eroded.

Remember that an even greater "dying" happened 250 My ago at the end of the Permian Period and the beginning of the Triassic. This marked a still greater rate of sudden extinctions of species than the K-T transition. An intriguing possibility, therefore, is that the Great Dying was also caused by an impact. The crater has probably been destroyed by plate tectonic motions. The issue remains a great puzzle, however. The problem is that no one has been able to find marked concentrations of iridium or other asteroid chemicals or any other signs of high-energy impact in the soil layer at the Permian-Triassic boundary. Perhaps it was a Permian-Triassic impact of an asteroid whose chemistry was too Earth-like for us to distinguish from Earth rocks; or perhaps it was a comet nucleus, whose icy material simply vaporized and merged with Earth's inventory of water. There is some evidence that the Permian-Triassic extinctions involved two stages, perhaps caused by some other non-impact process not yet understood.

LESSER IMPACTS, LESSER EXTINCTIONS

We cannot overemphasize that larger asteroids and craters are much rarer than small ones. (Of course, many fist-sized meteorites hit Earth every year!) Thus, regardless of the number of K-T-scale events that have happened, we would strongly predict additional impacts smaller than the K-T event, but big enough to have influenced life's evolution. Because they were smaller than the K-T event, their effects are harder to identify, but a few bits of scanty data of this sort have begun to come to light.

In 1982 the prominent geochemist R. Ganapathy found an iridium-enriched layer of sediment from near the end of the Eocene Epoch (a subdivision of the Tertiary Period), about 34 My ago, associated with the extinction of about five species of Radiolaria (tiny marine protozoans), along with rapid changes in general flora and fauna, including mammals. This layer of sediment is also associated with glassy spherules called tektites, another hint of a large impact. The crater itself has not been found. An ocean-floor drill core off New Jersey produced the thickest known layer of ejecta associated with late Eocene extinctions, and some researchers

65 My ago. One thousand years after the beginning of the Tertiary, the giant crater caused by the impact has refilled with ocean water and the world below has embarked on its new evolutionary course. The double-ring crater pattern has been found on many impact craters of this size that survive on our sister planet, Venus.

think the fatal impact feature lies on the sea floor there. Others have suspected the 100-kilometer Popigai structure in the Soviet Union, because several crude age measurements using radioactive isotopes place its age near 34 to 39 My ago. However, these ages have large uncertainty, as mentioned earlier, and at least one other age measurement places it at about 66 My, very near the K-T boundary! Clearly, we need better data on this structure; it seems large enough to have caused some extinctions, and we need to know whether it really does coincide with known extinctions.

A less clear-cut case of impact extinction is offered by events at the end of the Triassic Period and by the eroded impact crater, Manicouagan, in Quebec, Canada. Today it is a ring-shaped lake 70 kilometers across, but careful geologic studies prove that originally it marked an impact crater close to 100 kilometers across. The crater has been well dated at 210 ± 4 My. This seems to match the timing of a wave of extinctions at the boundary between the Triassic and Jurassic Periods, when 28 percent of the then-living reptile families became extinct. Various geologists have speculated that the Manicouagan impact was the cause of those extinctions. In early 1991, several new findings were announced that affirm an impact origin for the T-J extinctions. First, David Bice, a former student of Walter Alvarez, found what appears to be highly shocked quartz grains—a near guarantee of impact ejecta—at the T-J boundary in Italy. Second, his colleague Cathryn Newton of Syracuse University found an abrupt disappearance of Triassic clams and snails at exactly the same layer, confirming that the impact and extinctions were concurrent. Third, Columbia University geologist Sarah Fowell, studying T-J transition lake-bed sediments in New Jersey, discovered that the T-J plant and animal extinctions occurred within the extremely short interval of 21,000 years—an interval much shorter than uniformitarian paleontologists usually assign to species changes driven solely by Darwinian natural selection.

By ranking extinctions in order of magnitude we can say something about the connection of scale of impact and scale of the resulting eco-disaster. University of Chicago researchers David Raup and John Sepkoski, Jr. (1986) tabulated extinctions among 11,800 genera of marine animals during the last 270 My. Matching their figures with craters and other data, we find that:

- During the largest known extinction, the Great Dying at the Permian-Triassic boundary 250 My ago, something killed 61 percent of the genera that existed at that time. Perhaps this "something" was an impact that made a now-vanished crater bigger than 180 kilometers across, but no evidence of an impact event has been found in the strata.
- About 49 percent of the then-existing marine animal genera were wiped out by formation of the 180- to 200-kilometer Yucatán crater (possibly with subordinate impacts) at the K-T boundary 65 My ago.
- About 43 percent of the then-existing marine animal genera disappeared at about the time of the formation of the 100-kilometer Manicouagan crater in Canada 210 My ago.
- Probably the next largest extinction was at the Jurassic-Cretaceous boundary, when about 30 percent of these genera died out. As in the case of the Great Dying, no firm evidence of impact has been found for this event.
- Several additional smaller extinction events have been identified. Evidence of impact *has* been found for one of these, the event in which about 18 percent of these genera died at the end of the Eocene Epoch 34 My ago. Possibly the 100-kilometer Popigai crater in the U.S.S.R. is involved, since its date falls roughly in this range, but better dating is needed.
- Canadian geologists in 1987 located a 45-kilometer crater 50 My old on the Atlantic sea floor, southeast of Nova Scotia. It is not associated with an extinction event. (Raup and Sepkoski's chart shows only about 8 percent extinctions then, in the "noise level.")

Smaller craters don't correlate with major extinctions. Most other known giant craters date from before the Permian, so far back in time that the fossil record is either nonexistent or too poor to discuss extinctions with confidence.

Thus the data so far indicate that impacts that form craters at least 100 kilometers across create such disastrous climate changes that they cause global extinctions of many genera and species, while smaller craters of, say, 50 kilometers diameter and less seem to be down in the "noise level" of evolution, no doubt causing temporary global disasters but not extinguishing large numbers of species or redirecting the course of evolution in a major way.

Unraveling the exact sequence of the K-T and other extinctions, identifying the exact cause of the Permian-Triassic extinctions, and measuring the role of other large and small impacts during life's history—these are all exciting challenges for Earth scientists of the next generation!

B O O K S I X

The Most Unusual One Percent of Geologic Time

65 My Ago to the Present

14. Modern Times

> ...the earth beneath my tires, the rolling and solid ground, just a ball of dust and money and business lunches, spinning through the formatted emptiness of Einstein's curved space.
>
> —*Edward Allen,* Straight Through the Night, *1989*

s the dust from the K-T cleared, the modern world was being born, in terms of both geology and biology. The Mesozoic Era, or middle era of life, had ended. The new era, which we are in today, is called the Cenozoic Era—the era of recent life. It has been the age of the mammals, including us.

Because the Cenozoic is the most recent era, rich in diverse lifeforms, it has a more abundant fossil record than a comparable time span from earlier eras. This leads to a peculiarity in geology texts: the last 65 My generally gets vastly more attention than the identical time span from any earlier era, say 300 to 235 My ago. In a 1963 textbook about historical geology, the Cenozoic got 23 percent of the book. By 1988, more of the early history of Earth had been filled in, and the Cenozoic Era was reduced to 15 percent of a text published in that year. These numbers give a very false impression of

50 My ago. The landscape and its plants would be familiar to a visitor from the present day, although some of the creatures he or she might meet would not. Here a mylodon, a giant ground sloth, lazily munches on leaves growing twelve feet above the ground.

the Cenozoic's relative duration. In fact, the Cenozoic represents only the last 1.4 percent of Earth's history! We can't stress too much that the life-filled world we know represents only an unusual, passing episode of Earth's story.

FAMILIAR CONTINENTS AT LAST

Imaginary time-traveling astronauts, gazing toward Earth from space 64 My ago, "soon" after the K-T impact, would have been able to recognize the continental landmasses, albeit with some moments of hesitation. Earth's geography, though scarred by a fresh, 200-kilometer crater, was assuming its modern pattern. The primordial continents of Pangaea and Laurasia and Gondwanaland had broken into pieces more or less recognizable by their shapes and relative positions. South America had broken off from Africa, and the southern Atlantic was widening. Africa itself, which had been farther south, now straddled the equator. North America, barely connected to South America, had broken off from Europe more recently, and a strangely constricted North Atlantic Ocean was now expanding.

Far to the south, Australia had broken off from Antarctica and was moving north into the Pacific, while Antarctica neared the South Pole. A similar small continent, India, had broken away from eastern Africa and was moving north across the Indian Ocean, soon to collide with southern Asia.

THE EMERGENCE OF FAMILIAR MOUNTAINS

Just as the familiar continents were taking shape in the last 65 My, so also many familiar mountain ranges of the world were emerging. Mountains do not last long. Existing ranges are not old, in terms of the planet's age. The typical major mountain range, depending on its location, experiences substantial erosion in only 50 to 200 My. Suppose you magically piled up a huge artificial mountain; rain and rivers, loosening one grain of soil at a time, would erode much of it away within 100 My. For this reason, most of the spectacular craggy mountain ranges of modern Earth date from within the Cenozoic Era.

They resulted from collisions of continental plates. The resulting compression caused initially flat-lying sediments to buckle, and the convex portions arched to altitudes of 6000 meters (20,000 feet) and more. Of course, even as the land was being elevated, running water and flowing glaciers of ice began to cut stream channels and carry sediments downhill. The net result is carving of the mountain mass into canyons and steep hillsides. V-shaped valleys were cut by rivers, U-shaped valleys by glaciers, and gothic spires of resistant rock were left by erosion of the surrounding rock. All these are forms associated with young mountains. They exist only on Earth: other planets lack running water to erode uplifted masses. Lunar mountains are mostly impact crater rims, not remnants of erosion.

Earth's older mountain ranges have soft, rounded forms because their rocky spines have been worn down by erosion and their lower flanks have been partially buried under sediments. The Appalachian Mountains of the eastern United States are good examples. They were formed around 250 to 350 My ago as North America jostled against Europe in the collisions that produced Laurasia. In our era, they are eroding away, and only their rounded remnants are left.

A far cry from the Appalachians are the rough ranges produced more recently. The Rockies are an example. They rose around 65 My ago as North

40 My ago. The deciduous forests that are so familiar to Euro-American culture and mythology are a fairly recent addition to Earth, having dominated many continental surfaces primarily since the end of the Cretaceous Period.

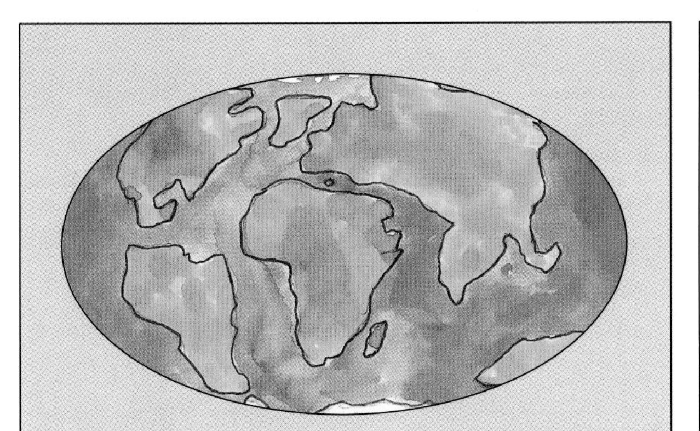

About 50 My ago, India crashed into the region of Tibet, crumpling parts of Asia to form the Himalayan mountains. Other continents were near their current configurations.

America shifted west, pushing into the Pacific tectonic plate and crumpling the western part of the continent. During most of the Cretaceous Period, the region now occupied by the Rocky Mountains of the western United States had been covered by a broad shallow sea that has yielded important fossils of Mesozoic marine lifeforms. But now this region was lifted and fractured. Just as Leonardo da Vinci pointed out in Italy, sea-bed fossils were raised to positions atop a high plateau. Erosion and further fracturing cut the high plateau into the ridges and valleys familiar from Colorado to Nevada today. This conversion of shallow sea to mountain range in North America has been called the Larimide Revolution. Before the discovery of plate tectonics, it seemed a mysterious, spontaneous convulsion of part of the inner Earth; now we see it as part of the larger pattern of continual crustal shifting driven by the thermal currents of the upper mantle.

In a similar way, the westward drift of South America produced the Andes, as mentioned in Chapter 11.

As India crashed into the region of Tibet some 50 My ago, another episode of crumpling produced the youthful range that now stands above all others, the Himalayas, crowned by Mount Everest at 8848 meters (29,028 feet) altitude above sea level. Crustal crumpling and thickening, and resultant mountain building, continued there until roughly 10 My ago. Thus we see that the Himalayas do not owe their grandeur to the magnitude of the plates that collided; India is small. Rather, they are grand because they are young. In 300 My, Mount Everest will be as worn down as one of the ridges in the Appalachians.

The Alps in Europe were also produced by compression, but did not arch as high as the Himalayas because Earth's crust was not as thick in that area. In fact, the Alps and Himalayas are linked, both being products of plate collision and jostling along the southern and western margins of Asia. Geologists have called them the Alpine-Himalayan Mountain Belt. During the last 5 My the Alpine region has been uplifted further, draining several shallow seas that had occupied parts of Europe.

Other well-known mountains built in the last 65 My include the Pyrenees of Spain and the Atlas Mountains of northwest Africa. The Pyrenees resulted from compression as Spain shifted toward Europe by about 50 kilometers, approximately 40 My ago.

Mighty mountain ranges from earlier eras, from the first three-fourths of Earth's history, were long gone by the beginning of the Cenozoic. We can only guess at their past grandeur by finding contorted strata that mark their crumpled cores.

Although some basic continental shapes were established by 65 My ago, the details continued to be refined throughout the Cenozoic Era. There were many local curiosities. Italy, for example, had not yet established its present position. Paleomagnetic data show that the Italian landmass rotated in orientation by more than 40 degrees between 65 and 50 My ago, and then rammed into Europe around 35 My ago, helping to build the Alps. It is easy to forget that such shiftings continue today and will continue into the future. Though movements may be fast on a geologic time scale, they are too slow for us to perceive. An active region is revealed only by the high frequency of earthquakes in the zones where plates are pulling apart or pushing together.

THE REPTILE/MAMMAL TRANSITION: WHAT THE PLATYPUS TELLS US

The vigorous emergence of mammals after the K-T impact may have been due as much to the removal of the dinosaurs as to the innate superiority of the mammals themselves. After all, mammals and dinosaurs had coexisted for a long time. Mammalian predecessors go back even farther. Mammal-like reptiles first appeared along with other early reptiles as early as the Permian, as much as 260 My ago. Somewhat crocodile-like, they were typically two to fifteen feet long and were distinguished by a sprawling posture, with sideward-extending legs and a larger, more ungainly head than some of the other reptiles. The mammal-like reptiles, along with some other reptile species, made it through the Great Dying at the end of the Permian 250 My ago. However, in the following period, the Triassic, they were left in seeming obscurity as the larger reptiles rose to glory. They may have been smaller, but they were nonetheless important, because by 220 My ago they had produced the first small, true mammals just as the dinosaurs themselves were emerging.

Mammal-like reptiles may seem like an evolutionist's desperate attempt to find missing links between reptiles and mammals, but we know something about them because certain transitional species still survive. For example, two curious Australian animals are classed as mammals but have a striking mixture of mammalian and reptilian traits. The duck-billed platypus and the spiny anteater are warm-blooded, have mammary glands, and suckle their young, like mammals. But unlike other mammals, they lay eggs, and they have the unusual sprawling limb structure of the early mammal-like reptiles. Only after the eggs are hatched do they care for their young in the normal mammalian fashion. They hint at the long-lost animals that made the transition from mammal-like reptiles to true mammals, as long as 220 My ago.

The early true mammals remained small and unobtrusive throughout the Mesozoic, from 220 to 65 My ago. Most were the size of a mouse, and few were larger than a house cat.

According to some interpretations, most present-day mammals descended from a relatively minor insect-eating variety of these Triassic mammals. Of course, in the Triassic Period, dinosaurs seemed to be the winners in the evolutionary saga: they had evolved a more agile, two-legged posture that may have helped them dominate in the competition for food and space. But we know what was to come. When the K-T impact ended the Mesozoic, and climatic changes devastated links in the food chain, the dinosaurs at the top of the food chain were among the hardest hit.

Without the impact, some scientists speculate, reptiles might have continued to evolve and might have produced an intelligent reptilian species. "We" might have been scaly, egg-laying creatures living in strange reptilian cities. With a little help from the K-T asteroid, mammals flowered instead, and gave rise to a species seemingly ill equipped to conquer the world and space around it. Soft-skinned, sparse in fur, lacking in horns, claws, fangs, or other body protection, we are weaker and slower than many of our fellow creatures, but we are blessed and cursed with a unique product of evolution: brains that allow us to design and analyze.

The most attractive challenge of this period is to trace how the previously obscure mammals proliferated to fill the environmental niche left by the dinosaurs after the K-T catastrophe. To reconstruct the exact sequence of biological changes, or to explain exactly why the mammals prospered in the face of adversity, is hard. We know from studies of the aftermath of the Mount St. Helens volcanic explosion that the biological effects of even local disruption can be amazingly complex. In a sudden catastrophe, factors such as season and time of day can make big differences to the pattern of extinctions and survivals. Consider two examples from the Mount St. Helens explosion, which happened at 8:32 A.M. on May 18, 1980. The eruption occurred a few weeks before pine buds opened to create new needles on the ends of branches. Many trees were smothered in ash, but the buds survived and opened, allowing many trees to live on. Had the blast come a few weeks later, all the needles would have been exposed and smothered, killing the trees. Second, animals that happened to be in their underground burrows at 8:32 A.M. survived the firestorm and crawled out through the ash blanket, while animals on the surface in the same regions were killed.

In the same way, season, species behavior patterns, and other factors may have been important in determining survival patterns after the K-T event. For example, many small burrowing mammals may have been protected from the K-T celestial firestorm, while the populations of larger dinosaurs, foraging on the surface, would have been exposed. These shy insect-grubbers may have been able to survive the ensuing months of darkness better than the reptile monsters who survived the holocaust. At any rate, the K-T event teaches us that macho doesn't always win out over modest. At least once, the meek *did* inherit the Earth.

WHAT KANGAROOS TELL US

Nature tried two experiments with mammalian development. Both involved keeping the embryonic young within the mother's body, instead of placing it in a vulnerable egg in an exposed nest. In the first experiment, the young are born alive at a very premature stage of development and crawl by instinct up the mother's body into a pouch for the rest of their growth. The animals who adopted this plan are the *marsupial mammals*. In the second experiment, the young are withheld in the mother's body for even more growth, being born as fully developed as possible (at relatively large size and consequently in a painful process). These are the *placental mammals*.

The placental mammals were more successful, making up 95 percent of Cenozoic mammals. The marsupial mammals, reminiscent of the reptile/mammal transition species, survive in Australia and South America. Australia's kangaroo is the most famous example. Placental mammals never became well established in Australia before humans arrived, so marsupials pursued an uninhibited course of evolution there.

ANIMAL DISTRIBUTIONS AND PLATE TECTONICS: WHAT WOLVES AND ARMADILLOS TELL US

One of the most interesting clues about the intertwining of geological evolution and biological evolution is so familiar that we miss its significance. Why is it that the continents of the Northern Hemisphere share many of the same animals, such as wolves, rabbits, and bears, while in the Southern Hemisphere, South America and Africa have markedly different species, and the animal life of Australia is even stranger, with its kangaroos, wombats, and platypuses?

The answer lies directly in the time scale of plate tectonics as compared with the time scale of biological evolution. Europe and Asia have been connected from some 250 My ago to the present. Major species have migrated and mixed throughout the entire area. North America had intermittent land bridges to Asia across the Bering Strait as recently as 12,000 years ago, and many creatures, from the now-extinct woolly mammoths to bears and wolves, roamed both continents.

In the south, however, continental isolation has been the rule for intervals longer than those required to make new species. The continents of the Southern Hemisphere were usually separated from Asia by the Tethyan Sea. Then South America and Africa broke apart some 100 My ago, leaving 100 My for relatively independent evolution. Prior to that time they shared the same fossils, but since then, species have been diverging. South America was isolated long enough to produce some unusual-looking forms such as armadillos, but eventually South America formed a land bridge with North America. New placental mammals from North America poured in, and many of the unusual South American species became extinct.

Australia has been the most isolated of all. It was part of Antarctica during the period of Gondwanaland and Pangaea, some 300 My ago. But the two pulled away from Africa, and India also broke off, isolating Australia/Antarctica by about 150 My ago. Antarctica always lay closer to the South Pole during this period, limiting its biological diversity. Finally Australia broke off on its own around 50 My ago. This history means that Australia has had 150 My of biological evolution independent of the other well-populated continents. The consequences for Australia, in terms of unique species, shocked the first European explorers, who arrived 150 My later and did not know what to make of beasts such as the duck-billed platypus.

The implications of independent biological evolution on the southern continents are far-reaching. We can say that the first 3000 My of evolution contained many episodes of mixing, so species were constantly "stirred," both in the seas and later on land. Of course, at any given time, there may have been one or two continents that had been isolated for 100 or 200 My, but subsequent plate tectonic collisions eventually brought them into contact with other continents, restirring the great global soup of life. This process may have injected once-localized creatures from small continents into more global settings, helping to explain the sudden flourishings of certain species. This in turn helps explain sudden changes in the fossil records of some individual continents.

With plate tectonics nature has carried out a nice experiment in evolution before our very eyes. In general, we learn that a few My is enough to produce dramatic new species, and 50 to 100 My is plenty of time for evolution to produce a great variety of whole populations.

Physicist W. G. Pollard, writing in 1979, used these circumstances to speculate about extraterrestrial evolution. He thinks of Australia as Earth-like

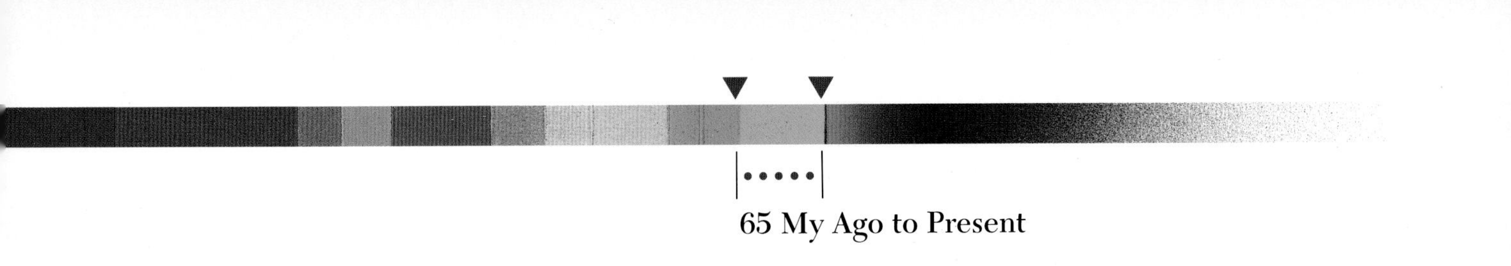

6 My ago. Massive waterfalls at the mouth of the Mediterranean formed as the Atlantic breached the Strait of Gibraltar after an era of Mediterranean evaporation. Gibraltar is on the skyline.

planet A, with an independent evolution for about 150 My. Similarly, he calls South America Earth-like planet S, with independent evolution for 100 My. The rest of Earth, he calls planet E. His point is that on these three "independent planets" evolution has *diverged*, with marsupials evolving only on planets S and A, and humans only on E. While some biologists have speculated that alien lifeforms would evolve along lines similar to our own, Pollard concludes that life (especially intelligent life) on planets beyond our solar system may be far more unfamiliar to us than the platypuses were to the first Europeans.

A GEOLOGICAL COINCIDENCE: A MATCHED PAIR OF SEAS

As biological evolution raced forward during the last 65 My, geological evolution continued apace. We tend to forget this. Perhaps the reader has secretly been wishing to have lived during some period when continents were being torn apart, when Earth might have seemed more spectacularly active. The truth is, we *do* live in such a period. These processes are happening all the time. Two familiar seas in different parts of the world are examples.

The Red Sea and the Gulf of California are similar in size and shape: long, straight, and narrow. There are reasons for the similarity. Both seas were produced at about the same time by a similar process, one called rifting. As continental blocks pull apart, depressed *rift zones* are often created. If the sea has access to the zone, water moves in as fast as the ground drops. The sea floor widens as the continental blocks separate. Fractures reach to a great depth through the crust, and volcanic magmas reach the surface, often producing a mid-ocean ridge as the center of the sea-floor spreading process. Both the Red Sea and the Gulf of California show these features.

The Red Sea began to open about 20 to 30 My ago, and after some on-again, off-again motions, continued to spread until less than 2 My ago. The Gulf of California has a similar history. Baja California began to break off from the Mexican coast, sliding in a northwestward direction, roughly 20 My ago. After a less active period, Baja California continued northwestward, widening the Gulf during the last 5 My. Much of this movement is along the notorious San Andreas fault, which causes many of the earthquakes in the vicinity of Los Angeles. Baja California and the coast of southern California are gradually being dragged into the Pacific. Studies of the San Andreas fault give evidence of 240 kilometers (130 miles) of shifting along the fault line in the last 5 My. Contrary to New Age superstition, southern California won't break off into the Pacific all at once in a mighty cataclysm, but will continue shifting northwestward earthquake by earthquake, a bit at a time, for millions of years to come.

The depression that forms the Gulf of California does not end at the northern end of the Gulf, but extends landward in a depressed zone. The lowland that was flooded by the Colorado River in 1905–06, creating the Salton Sea southeast of Los Angeles, is just a northward extension of the depression that forms the Gulf.

Both the Red Sea and the Gulf of California are rimmed by volcanoes, some active. They attest to the deep fracturing of the crust, tapping upper mantle magmas. Earthquakes continue to shake both regions, testifying to their continuing activity.

A GEOLOGIC CURIOSITY: A DRY MEDITERRANEAN SEA?

As we remarked in Chapter 11, the Mediterranean was originally part of the Tethyan Sea that stretched between Africa and Eurasia during the Cretaceous around 100 My ago. As the southern landmasses jostled against Europe and Asia, a curious accident happened. With the splitting of the Red Sea, about 20 My ago, Saudi Arabia moved north against Asia, closing part of the Tethyan Sea and sealing off what we now call the east end of the Mediterranean. Fossil evidence indicates that by about 6½ My ago, Africa and Spain touched and a narrow neck of land connected them across the current Strait of Gibraltar, allowing migration of common animal populations on both sides. Probably the land bridge was exposed when glaciation tied up much seawater in the form of ice, lowering the sea level. The land bridge temporarily sealed off the western end of the Mediterranean.

As the temporarily landlocked Mediterranean

Sea of 6 My ago basked in the sun, evaporation began. The sea level dropped. The remaining water grew saltier and saltier, because evaporation removes the water molecules, leaving the salts behind in solution. To give a modern example of the same process, the Great Salt Lake's salt content is high due to evaporation. Apparently this led to a *complete evaporation* of the whole Mediterranean Sea, probably within about 1000 years, creating a vast, salt-encrusted desert basin as much as 3000 meters (10,000 feet) below sea level—a basin that dwarfs the record now held by the Dead Sea Valley (392 meters, or 1286 feet, below sea level). Among the lines of geologic evidence for this event are:

- A million cubic kilometers of salt deposited during a 1-My interval on the floor of the present Mediterranean, including the deepest regions.
- Other minerals on the deepest sea floor of a type formed at shorelines of evaporating bodies of water.
- Canyons up to 1000 meters (3000 feet) below sea level of a type typical of land surface erosion.
- Structures on deep-sea floors resembling fossil stromatolites (see Chapter 9), which normally grow only in waters less than 50 meters (150 feet) deep.
- The present-day Mediterranean evaporation rate, which is three times the water-replacement rate by rivers and rain; only inflow from the Atlantic keeps the Mediterranean full today.

The most dramatic part of this story is not that the Mediterranean was once dry, but that it filled again. The Gibraltar land bridge could never have been very wide; probably when water breached this narrow neck, it eroded rapidly, allowing a series of rapidly changing waterfalls. Data from drill holes suggest that the cycle of evaporation and refilling happened at numerous times between 6½ and 5 My ago. Each refilling must have produced spectacular scenery. Gigantic waterfalls may have formed at the 2000-meter drop from the Gates of Hercules into the Mediterranean basin. It is unlikely that a 2000-meter-high single waterfall developed, yet who knows what geological wonders existed during these episodes? The waterfall, or series of waterfalls, delivered seawater at estimated rates of 1000 to 10,000 times that of Niagara Falls!

WHAT GLACIAL DEBRIS TELLS US

It is time to remind ourselves once again that Earth's climate has changed throughout geologic history. Today glaciers cover 10 percent of Earth's surface. During much of geologic time Earth's ice caps have probably been this size or smaller. But at times, as we have already seen, Earth underwent much colder periods. Glaciers occasionally covered much larger areas.

One of these glaciated periods included the last 2 My, when glaciers sporadically covered as much as 30 percent of the planet. During this period, ice sheets up to 4000 meters (13,000 feet) thick occasionally covered a great part of Canada and the United States, Greenland, Scandinavia, the U.S.S.R., and Antarctica. So much ice was tied up in these glaciers that the ocean level was up to 200 meters (650 feet) lower.

The recognition of the existence of ice ages was an interesting chapter in the development of geology, occurring in the mid-1800s. Already in the 1820s, geologists were perplexed by fossils of reindeer in southern France. Yet with the triumph of uniformitarianism over catastrophism, geologists could hardly imagine past climates that could have sustained reindeer on the Riviera. Another paradox arose from so-called erratic boulders that had been transported large distances from the outcrops of rocks where they originated. Elaborate theories of rock transport by flooding were developed to explain the erratic boulders, but further studies showed that floods did not produce the desired effects. Debates on these issues produced more heat than light.

The solution came from the most glaciated European country, Switzerland. An amusing story was told in 1835 by one of the participants in the debate, the French naturalist Jean de Charpentier. Walking in Switzerland with an uneducated woodcutter, Charpentier remarked on the huge erratic boulders. The woodcutter replied matter-of-factly:

> "There are many stones of that kind around here, but they come from far away, from the Grimsel [Mountains], because they consist of geisberger [granite] and the mountains of this vicinity are not made of it."
>
> When I asked him how he thought that these stones had reached their location, [reports Charpentier], he answered without hesitation: "The Grimsel glacier transported them.... That glacier extended in the past as far as the town of Bern."

In other words, Swiss woodcutters already knew what the European academics were struggling to find out: glaciers carry large rocks and, as the ice melts, dump them far from their source; and Europe had been covered by much larger glaciers, explaining the many erratic boulders. Charpentier's 1835 work was the first to explain various scratched bedrock surfaces as the result of ancient glaciers dragging sharp boulders across the landscape, instead of appealing to ancient floods or other agents. The reaction to Charpentier's theory was originally hostile (see A. Hallam's book in the references). However, the idea was soon taken up by the famous naturalist Louis Agassiz (1807–1873), and the idea of extensive ancient glaciation was well established by the 1850s.

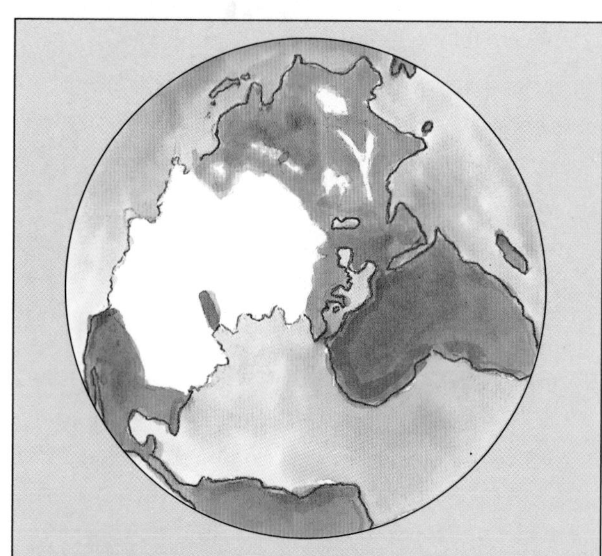

Glaciers spread over wide masses of Earth sporadically during the ice ages of the last few million years, including the last ice age, peaking about 18,000 years ago. As shown here, the polar ice fields expanded to cover much of the northern United States and northern Europe during such episodes of glaciation.

Deposits of glacial debris tell us that ice often spread over Europe, Canada, and the northern United States. Why? The answer is not clear. The major waves of glaciation came several hundred thousand years apart, but waxings and wanings of the glaciers occurred on even shorter time scales. There is some evidence for the beginning of a widespread temperature decline as long as 70 My ago. The ice ages of the last few My might be the culmination of this long-term climatic change, when the temperature finally dropped below a critical threshold for ice buildup. Numerous theories have been proposed to explain the quasi-periodic episodes of glaciation, but none seem to explain everything.

The most famous and far-reaching theory was a study by the Yugoslav astronomer Milutin Milankovitch, who in the 1930s calculated slight changes in the total sunlight reaching Earth at different times, due to slight variations in properties of Earth's orbit and axial tilt. For example, Earth's orbit is not quite circular: Earth is closer to the sun during a few months on one side of the orbit, and farther from it on the other side. More important, the degree of noncircularity changes over the millennia, so that the intensity of the seasonal variations shifts over thousands of years. Other properties of the orbit also change with different periodicities. Sometimes changes that act toward an average cooling are canceled by other changes that act toward warming. But sometimes the effects

6 My ago. Massive salt deposits (foreground) covered the floor of the Mediterranean but were flooded as waters of the Atlantic cascaded into the basin.

all add together to cause extreme cooling or warming. Milankovitch proposed that these quasi-periodic variations were a major influence on Earth's climate, and that the extreme cooling episodes could explain the ice ages. Certain measurements of ancient sediments reveal climatic temperature variations that match fairly well with the calculated Milankovitch effects, and therefore the Milankovitch cycles are probably at least part of the explanation of the ice ages. However, many geologists believe Milankovitch's astronomical cycles do not explain all the data. (A good review of the problem is a *Scientific American* article by Curt Covey; see the references.)

Subsidiary effects may have worked in concert with the Milankovitch cycles to produce the ice ages. These include continental drift, which shifted ocean currents and cooled some regions; continental uplift, which caused climates to cool and glaciers to spread; and changes in the carbon dioxide content of the atmosphere, which altered the greenhouse effect.

AN UNUSUAL CREATURE ARRIVES ON THE SCENE

During the last round of ice ages, in the last few million years, an unusual mammal evolved. All animals and plants had continued to evolve throughout the Cenozoic Era, and mammals were not to be left behind. Within about 10 My after the K-T impact, mammal evolution produced the first primates, initially an order of small, omnivorous tree dwellers distinguished by several new traits: an

1 My ago. Glaciers covered much of the northern United States during repeated cycles of glaciation during the last million years. The most recent cycle ended only about 12,000 years ago.

opposable thumb; wide-field stereoscopic vision; the birth of consistently smaller litters; and longer nurturing of offspring. The first trait allows manual dexterity, which would ultimately allow manipulation of sticks, stones, and the first tools. The second trait ensures accurate estimates of distance, as well as confident movement through tree branches. Fossils of the first primates include a mouse-sized, saucer-eyed primitive tarsier, whose bones were found in 50-My-old Wyoming volcanic ash beds in the 1980s. By about 50 to 45 My ago, the traits of an opposable thumb and stereoscopic vision were well established in various mouse-sized to cat-sized tree-dwelling primates. These were not monkeys or apes, but groups such as lemurs, with a more primitive skeletal structure. (The lemurs survive today only in Madagascar, because Madagascar broke off from Africa about when lemurs evolved but before more advanced predators appeared.)

By about 35 My ago, true monkeys and apes had evolved from these early smaller primates. There are three groups. One group developed independently in South America and is known as the New World monkeys. A different group, never in contact with the first, developed in Asia and Africa and is known as the Old World monkeys. The third group includes only five surviving examples—gibbons, orangutans, gorillas, chimpanzees, and humans; these are called hominids. Bone structures, DNA patterns, and gene-sequence experiments suggest that humans are most closely related to two animals in this group: the chimps and gorillas.

The earliest of the humans can be traced through scattered fossils, which are of course among the most exciting finds of paleontologists and anthropologists. Human precursors seem to have emerged in Africa some 4 My to 3 My ago. Fossils found by Louis, Mary, and Richard Leakey, Raymond Dart, and other famous researchers have revealed several varieties of chimpanzee-like creatures who stood about 4 to 5 feet tall and had limb bones nearly identical to ours but a brain with only half the volume of the modern human brain. By 2.5 My ago, these australopithecines were probably using both bone and crude stone tools. The australopithecines were at home on the ground and walked upright. This behavior may have been fostered by climate. Cooling conditions and uplift caused much of the African rain forest to be replaced by open grasslands, where upright posture on the ground was advantageous for seeing across the open savannas.

By about 2.5 My ago, one line of the australopithecines had apparently evolved to produce a more advanced creature, *Homo habilis* ("dextrous man"). *Homo habilis* made substantial tools. Recognizable choppers, scrapers, and other crudely shaped tools are found with these creatures from about 1.8 to 1.0 My ago.

A still more advanced line, *Homo erectus* ("erect man"), started as early as 1.6 My ago and died out as recently as 0.3 My ago (300,000 years ago). *Homo erectus* spread from Africa into Europe and Asia, reaching as far as China. Somewhere around this time, "real" humans were being born. It is hard to find the dividing line that defines "real" humans. Geologist Preston Cloud quips that if *Homo erectus* boarded a modern bus, the driver might find him acceptable but clumsy, while if *Homo habilis* tried to get on, the driver might call the zoo.

Homo sapiens ("wise man")—modern humans—finally evolved about 0.35 My ago, or 350,000 years ago. A variant known as Neanderthal roamed Europe, the Near East, and Africa until only 35,000 years ago. Although there is a popular image of Neanderthals as shambling dullards, their brains averaged some 10 percent *larger* than ours. They were stocky, with larger jaws and eyebrow ridges than those common among modern humans. They made a complex assemblage of tools including beautiful spear points, hand axes, and bone needles.

Humans indistinguishable from us (*Homo sapiens sapiens*) roamed Africa before 50,000 years ago (only 0.05 My ago, to express it in our favorite unit), and were painting beautiful animal portraits on their cave walls as early as 30,000 years ago. In less time than it takes to say *Homo sapiens sapiens*, they were busy getting into wars and writing books about the history of Earth.

THE HUMAN TIME SCALE

I have been deliberately brief with the history of humans, not only because it has been widely covered elsewhere in much more detail than we can offer here, but also, and more important, because humans represent only the last instant of geologic time. Human evolution gets a lot of attention, but it is just the latest flowering among a million flowerings in Earth's garden.

Furthermore, we must remind ourselves yet again that Earth's history does not end with humans or guarantee that we will remain unchanging overlords throughout the rest of time. Regardless of whether we manage to deal with our current problems or cause a global eco-disaster, Earth's history and biochemistry will go on. Whoever is on Earth a mere 10 My from now, if anyone, may not duplicate us any more than we duplicate chimpanzees. Living in the midst of the lightning-quick evolution of primates, we witness a veritable explosion of capability and intelligence, whether it be used for good or for evil.

Reflecting on the idea that human precursors have been around for 4 My, we realize with a shock that this represents a mere 0.1 percent of the history of Earth. Such a time interval is only a fraction of the duration of uncertainties that we have dealt with while discussing earlier lifeforms. For example, we remarked that the first appearance of land animals was pushed back in 1990 by fossil finds from the previous estimate of about 400 My to about 414 My. This "minor adjustment" of 14 My is nearly four times as long as the existence of our human-like ancestors on Earth. In other words, the duration of humanity thus far is so short that our whole species could be lost in the noise of geologic time.

To give humanity a space in this book proportional to our time on Earth, we (all of us since the australopithecines) should get no more than a short paragraph.

1 My ago. The so-called channeled scablands of the state of Washington bear witness to processes that must have occurred a number of times during the recent cycles of ice ages. Water confined by glacial ice was catastrophically released as ice dams melted and broke open. In Washington, massive erosion resulted.

1 My ago. Our ancestors roamed the African plains, using their newly developed stone tools to overcome the handicap of possessing few natural defenses. By organizing group hunts, these early communities were able to attack and kill large animals. Their diet was supplemented with plants, fish, and birds.

To put it another way, if we represent Earth's history by a one-day clock, starting at midnight, microscopic microbial life evolved at perhaps 4:00 A.M. Stromatolite algae colonies may have appeared by 6:00 A.M. The atmosphere changed from CO_2 to N_2O_2 around 1:30 P.M. The oldest substantial animal fossils of 700 My ago didn't appear till 8:00 P.M., and the well-recorded fossil sequence, starting with the Cambrian Period, did not begin until 9:10 P.M. Animals came onto the land at 9:45 P.M. and dinosaurs were thriving at 11:00 P.M. The K-T impact ended the reign of dinosaurs at 11:39 P.M. Our first australopithecine ancestors of 4 My ago appeared at 11:59 P.M.

Our civilization of the last few thousand years represents the pop of a flashbulb at midnight. The question is, What will be here in the first second after midnight?

1 My Ago to the Present

15. Earth Disasters: An Agent in Human History

. . . the cloud began to descend and cover the sea . . . most [of the crowd was] convinced that there were no gods at all, and that the final endless night of which we have heard had come upon the world. . . .

—Pliny the Younger, describing the eruption of Vesuvius, A.D. *79*

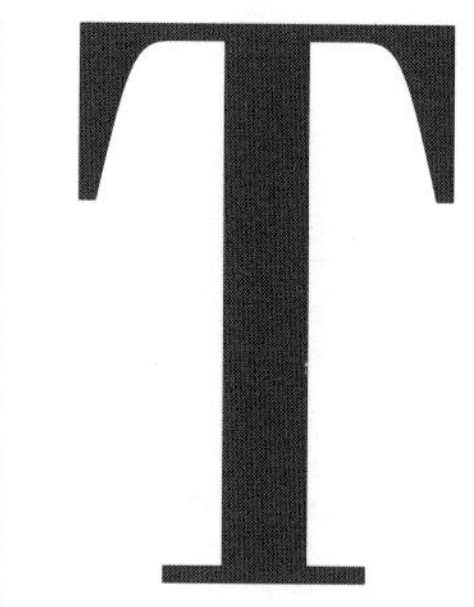

hroughout recorded history, the tiniest, most insignificant adjustments of Earth's crustal layers (as measured by planetary standards) have caused disasters for civilizations. The events that loom large by the standards of history are not necessarily an important part of Earth's story, and yet they are rich in human drama.

THE INNER WORKINGS OF VOLCANOES AND EARTHQUAKES

Why should massive, heavy, molten rock extrude from volcanic fissures, sometimes pouring out like molasses and sometimes shooting to heights of hundreds of feet? Lava has no motive force of its own. It flows only in response to other forces. What drives it upward through crustal rock and out of vents?

The answer lies in two related factors: the partly molten asthenosphere, or plastic zone, in the upper mantle; and the difference in density between molten lava and solid rock. Molten lava arises in the asthenosphere layer, some 100 to 250 kilometers below the surface, where upper-mantle heating and pressure conditions are right to create masses of melted rock. The molten conditions mean that the high pressure caused by the weight of overlying rock is distributed uniformly through the fluid magma. This means that if a fracture suddenly opens a clear channel from the asthenosphere to the surface, magma will be squeezed upward into the channel.

The second factor involves density. As rock melts, it expands slightly and acquires lower density than the same composition of solid rock. This difference in density causes a buoyant force that drives the molten rock upward. Imagine a vertical fracture from the asthenosphere to the surface, containing a vertical column of molten magma flush with the surface. Since the magma has a density lower than that of the surrounding rock, it weighs less than a similar-sized column of the rock. This means that the pressure caused by the magma's weight at the bottom of the magma column would be less than the surrounding pressure in the asthenosphere. As a result, it gets squirted upward. Think of a blob of oil in water. The oil has lower density than the water. The oil is analogous to the magma, and the water is analogous to the surrounding rock. If you release the oil under water, it rises. Of course, rock forms a stiffer barrier to the magma than water does to the blob of oil, but any fractures that appear in the rock allow the magma to rise.

This is essentially Archimedes' principle, named after its discoverer, the Greek naturalist of about 220 B.C. Any material of lower density in a deformable medium of higher density will tend to rise.

Archimedes' principle says that lower density magma will experience a buoyant force and, on average, try to ascend. If a vent opens in the ground and if the magma has access to the vent, it will be squirted outward, often with considerable force. This is aggravated by the fact that the magma contains dissolved gases, and as the magma ascends, the pressure decreases, gas bubbles form, and the magma becomes foamy. At the actual vent, the gas bubbles may drive the magma into the air at high speed, just as a bottle of soda or champagne fizzes when the pressure drops as the bottle is opened.

For many decades, the study of volcanoes was limited to describing the phenomena associated with their eruptions. Intrepid volcanologists established observatories on rims of seething craters, charting the magnitudes of lava outpourings and sometimes discovering rough periodicities from one eruption to the next. However, because of near-surface effects, such as the frothing of lava by dissolved gases, some of the most spectacular traits of volcanoes are merely superficial, having little to do with fundamental processes down in the earth. It is rather like trying to study the fundamental biology of human life by charting the eruption of pimples, instead of by studying DNA and blood circulation.

Only since the 1960s has it been possible to study volcanoes in a more fundamental sense. Arrays of seismometers and other devices have been set out across volcanoes such as Kilauea, in Hawaii Volcanoes National Park, Hawaii. These permit scientists to follow the movements of magma many miles under the surface as it is forced up from the asthenosphere and migrates from one set of fractures to another, causing earthquakes here and there.

Earthquakes themselves are monitored around the globe. As tectonic plates are pushed in one direction or another by sluggish currents in the mantle, rocks are stretched closer and closer toward their breaking point. When a rock unit fractures

3500 years ago. Eruption of the Thera volcano in the Mediterranean destroyed societies on several islands and may have been the origin of the legend of Atlantis.

underground, strong vibrations travel toward the surface, where an earthquake is felt. The underground site of the quake is called the *focus*, and the point on the surface above this, where the quake is most strongly felt, is called the *epicenter*.

The largest volcanic eruptions and earthquakes, though minor adjustments to the plates, wreak havoc among humans living nearby. Many legends and eyewitness accounts have come down to us of moments when Earth became restless. In this chapter we will describe some examples in chronological order.

THERA, CRETE, AND "ATLANTIS"—CA. 1500 B.C.

The modern crescent-shaped island of Thera (also called Santorin) lies in the Aegean Sea in the large triangle bounded by Greece, Turkey, and Egypt. This island and several of its smaller neighbors form the rim of an ancient volcanic crater roughly 12 miles across. Prior to about 1500 B.C., a larger volcanic island complex occupied this spot, possibly reaching 2000 feet or more above the sea. Dated volcanic debris testify that in roughly 1500 B.C. this volcano erupted in one of the larger known volcanic cataclysms, blasting vast quantities of ash into the sky and collapsing to form a giant caldera some 1300 feet deep.

This tremendous eruption was nearly forgotten for 3000 years. It might have been unimportant except that unlike most historic eruptions, it occurred in the very cradle of early Mediterranean civilization. Thera lies only about 70 miles north of Crete, home of the dominant civilization in the area in those pre-Greek times. The titanic explosion may have affected our cultural history in at least three ways. First, it may have caused the fall of Crete's civilization, thus allowing Greek culture to rise on its own; second, it may be the source of the myth of Atlantis; and third, it may relate to the punishments visited on Egypt, as described in the Bible's Book of Exodus.

Evidence of ancient disaster was found in 1860 when workmen discovered ancient ruins buried in volcanic pumice on one of the outlier islands in the Thera group. Later, archaeologists on Crete uncovered ruins of the great ancient city of Knossos, capital of a widespread culture. This culture was called Minoan by the archaeologists who unearthed it. Further excavations showed that the Minoans dominated much of the Mediterranean after 2500 B.C., but suddenly collapsed for unknown reasons around 1500 to 1450 B.C.

In 1939, Greek archaeologist Spyridon Marinatos revived an earlier theory that the cause of the Minoan collapse was the mighty eruption of Thera. Marinatos found evidence of seaborne pumice in ruins on the north coast of Crete. Pumice is a sponge-like form of lava, so light that it floats; Marinatos theorized that a giant tsunami wave had washed pumice inland and inundated coastal regions. A monster tsunami would surely have been generated if Thera collapsed all at once to create the deep caldera now enclosed by the ring of islands at modern-day Thera. Estimates have placed the height of the tsunami wave at 200 feet or more, but the estimates are questionable. A 210-foot tsunami—the highest on record—was observed on Kamchatka in 1737. Such waves can send water to even higher elevations in the narrow coastal valleys; one reached altitudes of 1720 feet in Alaska in 1948.*

Other effects of the Thera eruption damaged Crete. For example, studies of sea-floor drill cores in 1965 proved that the whole area, Crete included, had been blanketed by ash from Thera. The main ash deposits lie to the southeast, consistent with summertime wind patterns in the area. A summer ash fall could have damaged crops. In addition to the simple smothering effect, ash from some volcanoes contains fluorine and other chemicals that damage vegetation. Crete was also devastated by major earthquakes during the eruption. Even the smaller, more recent eruptions at Thera, recorded by naturalists, were accompanied by severe quakes in this region.

Curiously, recent archaeological evidence suggests that Minoan civilization persisted in Crete for about a generation after the eruption, but that cities, palaces, farms, and even rural shrines were then destroyed by fire, bringing about the final collapse of the society, around 1450 B.C. Many settlements were ransacked and not rebuilt. Artistic levels declined, and stylistic influences intruded from the less cultured Greek mainland, which lay to the northwest and was less affected by the eruption. The best interpretation of the data seems to be that Thera dealt the initial devastating blow to the Minoan civilization, with earthquakes destroying palaces, tsunamis wiping out many ports, and ash falls possibly damaging summer crops. Some of the Minoans left the area, while others tried to rebuild. Later, invaders from the Greek mainland wiped out the remaining Minoan society. Mainland Greek culture eventually prospered, culminating in the golden age of the Greeks nearly a thousand years later.

Have these long-ago events been carried down to us in myth? In the 1960s, archaeologists working on these problems revived a century-old idea that Crete and Thera were the source of Plato's "lost continent" of Atlantis. Plato, writing around 370 B.C., recorded information supposedly given to the Greek statesman Solon by Egyptian priests when Solon visited Egypt about a century earlier. According to the old legends, Atlantis had been an island nation that had ruled the Mediterranean thousands of years earlier, but Atlantis had been destroyed in a single day and night and sunk beneath the sea. This might be a distorted version of the disappearance of the original island of Thera, its ancient towns, and, in short order, Knossos and its Minoan colonies.

Provocatively, Plato's description includes curious remarks that the city of Atlantis was a port arranged in concentric circles around a central island,

*See the reference by Vitaliano for further details of tsunamis and a discussion of the Thera eruption.

which had both hot and cold springs—details that could fit the preeruption configuration of the giant volcanic crater that formed preeruption Thera.

A more speculative controversy about Thera is that the great eruption happened during the time the Israelites were captives in Egypt, about 500 miles to the southeast, and that it was the source of at least some of the Egyptian disasters recorded in Exodus:

> Then Moses stretched forth his rod toward heaven; and the Lord sent thunder and hail, and fire ran down to the earth . . . and there was thick darkness in all the land of Egypt three days.

Mapping of sea-floor ash deposits showed that the dense Thera ash cloud did travel southeast toward Egypt, supporting the possibility of a temporary darkening. Also, hailstones are a common part of ash falls, since ice condenses on ash grains. Finally, luminous electrical discharges were witnessed 200 miles from Hekla volcano, in Iceland, when it erupted in 1766. It has been suggested that all these phenomena were witnessed in Egypt, along with tsunami surges that may have been responsible for flooding, swamp formation, and the resultant swarms of frogs and flies mentioned in Exodus.

The main controversy about this interpretation centers on the dates: some authorities dispute whether or not the dates of the eruption and the exodus from Egypt match; some authorities place the exodus in the 1400s B.C., but most place it in the 1200s B.C. Perhaps further dating of the ash and future improvement of biblical chronology can settle this issue.

Some authors appeal to the Thera eruption to explain still other events recorded in famous Greek myths. (See Vitaliano's book in the references for an interesting summary.) The chances seem good that we have at least some cryptic cultural memories associated with this lethal episode from the dawn of Western civilization.

VESUVIUS AND POMPEII—A.D. 79

Vesuvius is a volcanic mountain located on the coast of the scenic Bay of Naples, in Italy. The beauty of the region has attracted settlers since ancient times. In Roman times, Vesuvius had been inactive for several centuries and was not considered dangerous. The famous slave Spartacus, who led a revolt in A.D. 73, hid for a while in Vesuvius's crater. On opposites sides of the mountain in A.D. 79 were the Roman towns of Pompeii and Herculaneum, where many fashionable Romans maintained summer homes. Pompeii had 20,000 inhabitants.

Among the inhabitants was Pliny, a famous naturalist and the commander of a local naval squadron; he was staying in his villa across the bay from Vesuvius with his sister and his eighteen-year-old nephew, known as Pliny the Younger. On August 24, Vesuvius sent out a great cloud, frightening local residents. Pliny set out across the bay to rescue friends at the foot of the mountain. He was never to return, and Pliny the Younger wrote an account of what happened during the next, fateful day. He reported (in a loose translation here) that by the next morning the cloud had spread. The region was being shaken by earthquakes:

> The light was very faint and doubtful. The buildings all around us tottered, and although we stood on open ground, . . . there was no remaining without imminent danger. We therefore resolved to leave the town.

Pliny the Younger must have been an astute observer of human nature as well as a good writer, because he noted the psychology of the crowd as people prepared to leave during the tremors:

> A panic-stricken crowd followed us, and (because to a mind distracted by terror every suggestion seems better than its own) they crowded in on us as we came out. When we had gotten away from

> the house, we stood still, in the midst of a most dangerous and dreadful scene. The chariots, which we had ordered prepared, were so agitated backwards and forward, even though on level ground, that we could not keep them steady, even by supporting them with large stones. The sea seemed to roll back on itself and to be driven from its banks by the convulsive motions of the Earth. The beach was considerably expanded and several sea animals were left upon it.
>
> On the other side of the bay, a black, dreadful cloud, broken with rapid zigzag flashes, revealed behind it variously shaped masses of flame.

This was a dense sheet of smoky debris, thick airborne ash, and mud rushing down the mountain toward the various settlements. Pliny the Younger reported from a distance what happened next as they headed down the road away from the angry mountain:

> Soon afterward, the cloud began to descend and cover the sea. It had already surrounded and concealed the Island of Capri. . . . A dense mist seemed to be following us. "Let us turn off the main road while we can still see," I said. "If we fall down here, we might be crushed to death by the crowds still following us."
>
> Scarcely had we sat down when night came upon us, not such as we have when the sky is cloudy or when there is no moon, but that of an enclosed room when the lights are out. You could hear the shrieks of women, the screams of children, and the shouts of men; some calling for their children, others for their parents, others for their husbands . . . ; one lamenting his own fate, another that of his family; some wishing to die from the very fear of dying; some lifting their hands to the gods, but most convinced that there were now no gods at all. . . . A heavy shower of ashes rained upon us, which we were obliged every now and then to stand up and shake off, otherwise we would have been crushed and buried in the heap.

In Pompeii, closer to the mountain, the eruptive debris swept across the town and thousands of people died. In Herculaneum, the people had enough time to evacuate, but the city was soon covered by as much as 65 feet of ash and lava.

The location of the two towns was more or less forgotten, and they were not rediscovered until 1738, when treasures began to be found by excavators. Pompeii, found in 1748, created a sensation and affected European neoclassical architecture. Archaeologists have located more than 2000 skeletons of people who died in the streets, often clutching cloths to their faces. They were asphyxiated by the thick volcanic gases that settled on the town. Many carried bags of coins or other possessions as they, like Pliny the Younger, tried to flee their homes. Many of the bodies created cavities in the ash flow, and when plaster was poured in the cavities, molds were made showing the bodies as they fell. These molds are visible today.

Pompeii and Herculaneum were not specifically described by Pliny the Younger, but were popularized in many books, including a novel, *The Last Days of Pompeii,* published in 1834. Though not the largest volcanic disaster in history, it was probably the most famous in European cultural history.

THE LISBON EARTHQUAKE—1755

November 1, 1755, was All Saints' Day in Lisbon, Portugal, and parishioners crowded churches in this prosperous, beautiful city of 300,000. At 9:30 A.M., just as priests were ascending the altar in the great basilica, the whole city was shaken by an earthquake whose epicenter was offshore. The walls of the church and other buildings began "to rock and sway like an unsteady ship." In the next half hour, three separate shocks collapsed various churches and thousands of other buildings. Many people were crushed inside, while others rushed into the streets, whereupon a 40-foot tsunami rushed across the town. An estimated 60,000 people died in the earthquake, the tsunami, and subsequent fires.

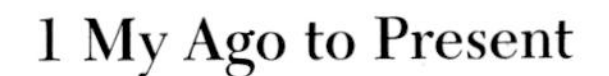

1900 years ago. The burial of the Roman city of Pompeii was one of the first great natural disasters in recorded history. The eruption, in the year 79, buried the city first under six feet of pumice and hot scoriae that collapsed roofs, then under a blanket of volcanic ash five to six feet deep. So thorough were the destruction and burial that the city lay virtually forgotten for nearly sixteen centuries.

There is a story that the king, who survived by being away from the palace, was approached by priests and asked to order prayers for forgiveness, under the theory that the quake was retribution for the sins of the people. The king turned to one of his nobles to ask for advice. The adviser he chose was the marquês de Pombal, who went down in history for looking forward instead of backward. De Pombal reportedly answered, "Sire, we must bury the dead and feed the living." He was thereupon given power to do this. He implemented his own advice, redesigned the city, and in the process collected hundreds of records of the event and the damages done, creating the first set of scientific records of such a disaster.

The Lisbon earthquake had a strong bearing on the development of European philosophy and science. Voltaire, in his satirical novel *Candide*, philosophized about a universe in which parishioners could be demolished in the act of worship. Catastrophist and Neptunian geologists referred to the Lisbon quake as evidence that mountains had been thrown up in sudden cataclysms. Today we see it as but one more in Earth's long series of adjustments.

THE YEAR WITHOUT SUMMER—1816

The largest known volcanic eruption in recorded history was probably the eruption of Tambora volcano on an island in Indonesia in 1815. It ranks first in magnitude out of 5564 historic eruptions tabulated by the Smithsonian Institution in 1981, and has been called (probably with less certainty) the most explosive eruption in the last 10,000 years. Considering its effects at the time, this event is curiously little known to the modern public. The reason is mainly that European naturalists and government agents were just getting a toehold in the southwestern Pacific, and communication was spotty. We know from comparisons of the mountain before and after the blast, as well as from ash deposition, that the volume of blasted material was the largest among documented eruptions: roughly ten times that in the Thera eruption or in the more well-known 1883 Krakatoa eruption, and roughly 100 times that of the A.D. 79 Vesuvius or the 1980 Mount St. Helens eruptions. The mountain was reduced in height by 4200 feet, and an estimated 25 cubic miles of ash and debris was ejected.

A week before the main explosion, blasts were heard many miles away. Some 450 miles away at a military post, a detachment of soldiers was dispatched to see if an attack was underway. After the main eruption of April 11, 1815, the lieutenant governor of Java gathered reports of the event from the whole region, and from these accounts, Charles Lyell was able to describe the blast in the first edition (1830) of his *Principles of Geology*:

Hartmann photo

Changing sunset colors are produced by the atmosphere's varying combinations of transparency, density, and dust and moisture content. Twin Peaks, San Francisco.

> The sound of the explosions was heard in Sumatra, at the distance of 970 geographical miles. . . . Out of a population of 12,000, only 26 individuals survived on the island. Violent whirlwinds carried up men, horses, cattle . . . into the air, tore up the largest trees by the roots, and covered the whole sea with floating timber. . . . So heavy was the fall of ashes, that [the ashes] broke into the Resident's house . . . 40 miles east of the volcano, and rendered it, as well as many other dwellings in the town, uninhabitable. . . . The floating cinders to the westward of Sumatra formed, on the 12th of April, a mass two feet thick and several miles in extent, through which ships with difficulty forced their way.
>
> The darkness occasioned in the daytime by the ashes in Java was so profound, that nothing equal to it was ever witnessed in the darkest night.

The ash pall made day as dark as night within 200 miles of Tambora for three days following the eruption.

The blast involved an energy expenditure estimated at three times the total nuclear weapon stockpiles of the 1980s.* Ash layers destroyed local crops, causing severe famine and illness. Although the eruption killed 12,000 on Tambora's island, an estimated 90,000 lives were lost in Indonesia generally, due to the wider effects.

But the effects were far from local in time or space. Dust from the eruptions was apparently ejected from the site, at about 7° south latitude, into the jet stream of the Northern Hemisphere, whence it spread over northern North America and Europe, reducing sunlight. The following, frigid year was called "the year without summer" by bewildered Europeans and Americans, who did not know that planetary events half a world away were affecting their lives.†

"The year without summer" produced snowfalls in June 1816 in New England, and temperatures there plunged below freezing during several mornings in July and August. That summer was at least as disastrous in Europe. July of 1816 was the coldest recorded in the Lancashire region of England in 192 years of records, beginning in the 1700s. Geneva experienced the coldest summer since 1753. Crops failed, initiating serious famine in some regions. In an extraordinary bit of economic history, the price of wheat shot up due to the intercontinental crop failures and reached a peak in 1816 that was never again surpassed until 1972–73, during the Russian wheat shortage and the Arab oil embargo.

Perhaps due to the vagaries of jet stream circulation, the cooling effects of the dust pall were mixed in other regions: famine in Bengal followed the summer of 1816, but Japan's rice yields were not exceptionally low. In China, spotty weather records, made without thermometers, indicate a cool, but not extreme, summer. Despite the incomplete weather records, most analysts ascribe the aberrant weather of "the year without summer" to the high-atmosphere dust pall from Tambora.

The link between Tambora and subsequent events is more speculative, but provocative data suggest that the eruption also led to the first worldwide epidemic of Asiatic cholera, which caused hundreds of thousands of additional deaths from 1816 to 1830. Asiatic cholera had previously been confined to regions involved in Hindu pilgrimages to the Ganges, although it occasionally spread into China. The famine in Bengal in 1816–17 led to another outbreak there. It spread along the Asian trade routes, possibly due to British military movements. By 1819, cholera was in south India, by 1823 in Turkey and Asia Minor, and by 1830 in Moscow. Some 50,000 people fled Moscow, and 200,000 people reportedly died in Pest, Poland. By 1832, cholera had spread into Europe and Egypt. Paris could not keep up with the burying of the dead; corpses were stored in public buildings. And in 1832 the disease showed up in New York City, where deaths approached 100 per day, mostly

*In the parlance of physicists, the energy yield was an estimated 1.5×10^{20} Joules. The results are discussed in a book by Stommel and Stommel (1983) and are compared with the results of a nuclear war by Soviet physicists Georgi Golitsyn and Aleksander Ginsburg in a 1985 article. See the references.

†If it seems unlikely that ash from a few degrees south of the equator would end up in the stratosphere over North America and Europe, note that such a result was definitely confirmed in 1963 when Mount Agung, a volcano near Tambora, erupted and produced about one-thousandth as much debris as Tambora; yet Agung's stratospheric ash was measured and produced vividly colored glows after sunset for some months in North America.

Although New England folk did not know what had hit them, one famous early American naturalist had already speculated about the volcanic haze causing the cold weather. Benjamin Franklin noted a high haze layer in the summer of 1783, followed by an unusually cold winter. Unknown to him, the haze was from a Japanese volcano. He speculated that the haze was from either a volcano or meteorite debris falling from space, and he wrote that sunlight's "summer effect in heating the earth was exceedingly diminished. Hence the surface was early frozen [and] the winter of 1783–84 was more severe than any that happened for many years."

among the poor, and again a backlog of corpses built up.

The biological consequences of this eruption and cooling, including hundreds of thousands of deaths, are a grim testimony to the possible effects of the large explosions that dot Earth's history, whether from volcanoes, asteroid impacts, or, in modern times, nuclear war. Our civilization is a delicate house of cards and can easily be disrupted.

KRAKATOA—1883

Sixty-eight years after Tambora erupted, another nearby volcano exploded, producing only a tenth the energy of Tambora's eruption but gaining much wider fame. Krakatoa's fame arose mainly from the fact that it was the first major eruption to occur in the era of worldwide communications and scientific instrumentation. Krakatoa was known in 1883 as an extinct volcano in the area of first landfall for ships sailing eastward across the Indian Ocean. It is located in the strait between the large islands of Java and Sumatra. As gateway to the East Indies, this strait was a target for many voyages. A 1983 anniversary volume released by the Smithsonian Institution cites many of the dramatic original reports (Simkin and Fiske, 1983).

In May 1883, eruptions began, and a local ship was sent to investigate from nearby Djakarta, Indonesia (then Batavia, Netherlands East Indies). This ship arrived on May 28, and a mining engineer on board noted that already most of the vegetation on the island had been destroyed. He wrote:

> From the middle of this dark and desolate landscape, the epitome of total destruction, a powerful column of smoke of indescribable beauty drifted over the sea. . . .

The sense of beauty lasted only three months. On August 26–27, 1883, catastrophe began. At 1:00 P.M. on August 26, Krakatoa blasted out enough ash to plunge the whole region into total darkness for three days. Another mining engineer in Djakarta noted the cooling effects of the cloud: the temperature leveled out at a full 12 degrees Fahrenheit below that of the previous days. The log of a nearby British ship described many of the phenomena we have encountered in the earlier accounts:

> . . . darkness spread over the sky and a hail of pumice stone fell on us. . . . The blinding fall of sand and stones, the intense blackness above and around us broken only by the incessant glare of varied kinds of lightning, and the continued explosive roars of Krakatoa, made our situation a truly awful one.

In such eruptions, such a large volume of material is blown out of subterranean magma chambers that cavities, or at least regions of suddenly lowered pressure, are produced underground. This leads to collapse of the volcano itself, as was the case with Krakatoa. The final major eruptions of Krakatoa began around midnight, culminating in the collapse of the mountain to form a giant undersea caldera after dawn on August 27. One ship's captain noted in his log that several of his crewmen were struck by lightning during the night, and by 2:00 A.M. ashes were lying three feet thick on the ship's deck. These events were so violent that their noise was reported at several sites 2000 miles away and at one site in the Indian Ocean nearly 3000 miles away. In South Australia, at the 2000-mile mark, people were awakened from their sleep by the rumble of the distant explosions.

Any large explosion sends a wave of pressure racing out through the air at the speed of sound. It dissipates in a matter of feet or miles, depending on the loudness of the sound. But the pressure wave caused by the largest Krakatoa explosion was recorded by sensitive barometers in England and elsewhere not just once but several times as the pressure wave traveled around the globe again and again until it got too weak to detect.

The greatest local disaster came in the morning

New land being created in a dramatic conjunction of the sea and molten lava. Kalapana, Hawaii.

from tsunamis generated by the collapse of the mountain into the sea. The waves killed an estimated 36,000 people in coastal towns. At around 11:00 A.M., a ship steaming out of the harbor of the town of Telukbetung met the wave head on. The ship's chief engineer reported that

> we were lifted up with a dizzy rapidity. The ship made a formidable leap, and immediately afterwards we felt as though we had been plunged into the abyss. . . . Immediately afterwards another three waves of colossal size appeared. And before our eyes this terrifying upheaval . . . consumed in one instant the ruin of the town, the lighthouse fell in one piece, and the whole town was swept away in one blow like a castle of cards. All was finished. There, where a few moments ago lived the town of Telukbetung, was nothing but the open sea.

About 5000 people perished in this town alone.

The eruptions subsided, but the effects lingered on around the world. As is typical of large volcanic eruptions, fine ash was lifted into the stratosphere and circulated around the world. Being so much higher than most dust, such ash catches the rays of the sun well after sunset, producing eerie glows. In 1883, unusual red and greenish sunsets were reported in North America and Europe for months after the eruption.

THE TUNGUSKA IMPACT EXPLOSION—1908

A mysterious explosion occurred over the Tunguska region in Siberia on June 30, 1908. This event was certainly not one of the great disasters of history, and might better be classed instead as one of the great curiosities. However, as we will see, it could be a harbinger of a much greater disaster in the near future. It is important precisely because it is not an effect of Earth's internal workings, but rather a random encounter with a piece of debris from Earth's larger, cosmic environment. The ominous thing about cosmic debris is that while the Tunguska object was moderately big, we will undoubtedly encounter even bigger objects if humanity waits long enough.

Like the air pressure wave of the Krakatoa explosion, the pressure wave of the Tunguska explosion was noted in English meteorological observatories 2200 miles away. Seismic vibrations were recorded 600 miles away, and witnesses 300 miles away reported "deafening bangs" and a fiery cloud stretching across the sky (see reports collected by Krinov, 1966).

As the meteorite descended through the sky, it was brilliant. Witnesses at a distance of about 130 miles reported "an irregularly shaped, brilliantly white, somewhat elongated mass . . . with [angular] diameter far greater than the moon's."

The blast was energetic. Carpenters were thrown from a building and crockery was knocked off shelves. An eyewitness about 45 miles from the blast reported:

> Suddenly . . . the sky was split in two and . . . above the forest the whole northern part of the sky appeared covered with fire. . . . I felt great heat as if my shirt had caught on fire. . . . There was a . . . mighty crash. . . . I was thrown onto the ground about 20 feet from the porch. . . . A hot wind, as from a cannon, blew past the huts from the north. . . . Many panes in the windows blew out and the iron hasp in the door of the barn broke.

The closest known witnesses were some reindeer herders asleep in their tents about 50 miles from ground zero. They and their tents were blown into the air and several of them lost consciousness momentarily. "Everything around was shrouded in smoke and fog from the burning, fallen trees," they reported.

These events were caused by a relatively small bit of interplanetary debris colliding with Earth. The debris may have been a relatively small piece of asteroidal rock, but was more likely a fragment of a comet, consisting of weak, carbonaceous rock with ice embedded in it, matching the composition of comets. The preimpact orbit in space has been estimated from reports of the meteorite's flight path. The shape of this orbit around the sun supports the idea that the object was cometary, not asteroidal.

Although the energy of the explosion was equivalent to a 10- to 15-megaton nuclear bomb, the object left no crater because it exploded in the upper atmosphere. Large rocky and metallic meteorites may burn off some material in the atmosphere, similar to a reentering spacecraft, but they usually reach the ground. The reason for the airburst in this case is that the ice in this object probably turned to steam from the heat of atmospheric entry, blowing the whole object to smithereens. From the energy of the explosion and other characteristics, calculations indicate that the object was about 20 to 60 meters in diameter (65 to 195 feet).

Because the event was a single high-altitude explosion, not a continuous eruption of ash, it did not turn the sky black for days. However, it did eject a lot of dust into the high atmosphere. For weeks after the event, sunsets were unusually colorful and lingering and the night sky in Europe and Russia was unusually bright, telltale indications of fine dust in the stratosphere.

Early expeditions to the site revealed that trees for miles around had been blown over. They all lay with their trunks pointing back at the site of ground zero. At ground zero, trees were standing, but their branches were stripped off by the blast that had occurred directly overhead. A 1961 expedition recovered tiny carbonaceous meteorite particles, proving that the object had indeed been interplanetary (contrary to recent pseudoscientific pop literature claiming the object to have been a nuclear-powered interstellar spaceship that exploded!).

Two decades ago, the Tunguska explosion would have been written off as a random, isolated event that had nothing to do with geology or threats to civilization. Now, however, we recognize that asteroid/comet impacts happen all the time, and that the smaller they are, the more often they occur. As we have seen, truly catastrophic events, like the K-T impact that killed off the dinosaurs, are hundreds of millions of years apart. Atom-bomb-sized explosions such as the Tunguska event may happen as often as every few centuries. Of course, more than three-fourths of them hit far out at sea or in the polar regions, which explains why there are few historic records.

Nonetheless, our cameras and other recording devices are picking up more and more objects in nearly this size range. On April 10, 1972, a house-sized interplanetary body fell at a shallow angle into the atmosphere over Utah, Idaho, and Montana. It dipped as low as 40 miles up, but by chance, due to its shallow angle of flight, it skipped back out into space, like a stone skipping off water. It was well photographed in movies and still pictures by tourists in the area of Yellowstone Park. Had it hit the ground, it would have caused an explosion approaching the magnitude of Hiroshima in Alberta, Canada. (See the article by Jacchia in the references.)

In 1991, telescopes picked up an asteroid in space that missed Earth but came closer than any other known such object, only about half the distance of the moon. It was very nearly the same size as the object that struck in Tunguska. From such observations, we know it is probably only a matter of chance and time—probably only a generation or so—until some celestial visitor heads toward another large explosion somewhere on Earth.

WHAT DISASTERS TELL US

The above may sound gloomy, but there are reasons for optimism about natural disasters of all sorts. First, there is a big difference between threats to civilization and threats to Earth as a whole. As the Tambora eruption showed, volcanoes are capable of significant disruptions of economics and health while being only negligible glitches from the point of view of the planet. Even if human civilization is disrupted for decades or more by modest-scale catastrophes, life will go on as long as the global ecosystem isn't destroyed. Internally caused disasters such as volcanoes, earthquakes, and plate tectonic processes are probably not a threat to the global ecosphere; they cause local destruction but aren't big enough to threaten planetary life. This statement requires a caveat: we still don't know what caused the biggest biological disaster, the Great Dying at the end of the Permian 250 My ago. In any case, we have reason to worry about the threat to life from interplanetary bodies. The K-T event 65 My ago showed life's resilience to even large catastrophes. But Earth as a whole is safe from any disasters we know about within the solar system.

Second, we learn from such disasters that life is not a sure thing for those who sit back passively, waiting to enjoy whatever nature brings. If we want to increase our chances for safety over the long term against disruption of the economy or the ecosphere by planetary catastrophes, we will want to pursue better ways of understanding and dealing with them.

Third, as a species, we are getting more capable of dealing with threats to civilization. We are learning to predict internally caused disasters. By studying earthquake signals and other signs of internal activity, geologists are beginning to have success in predicting eruptions and earthquakes. Adequate preparations reduce damage.

Asteroid/comet impacts are another matter, because the scale of such a disaster can be global, as in the K-T event. If we build a truly interplanetary civilization, with beachheads on several planets and in orbiting space cities, we can help ensure our survival in the event of a catastrophe to Earth's ecosphere. As will be seen in the next chapter, once we can operate freely in space, we will be able to do something about the hazards of possible large impacts in the next century.

BOOK SEVEN

Beyond Our Time

The Present to 1 My from Now

16. The Near Future

> Over the course of the day, and over the subsequent days, it became clear that the world was not ending any faster than usual. . . .
>
> —*Edward Allen, from* Straight Through the Night, *1989*

In terms of the twenty-four-hour clock, which represented the history of Earth in Chapter 14, the big problem facing planet Earth is what will happen in the first fraction of a second after midnight. Our location in time is midnight, the end of the first day, and the next million years represents only the first nineteen seconds after midnight. The next 500 years, which may decide the fate of humanity and set the precedent for how technological civilization treats its mother, Earth, represents only the first hundredth of a second after midnight.

That hundredth of a second will reveal whether or not humanity solves its survival problem. The survival problem can be stated quite simply. Do we remain restricted to planet Earth and allow our activities to continue altering the environment of the planet, thus threatening our own species as well as others? Or, alternatively, do we bring the planet-damaging aspects of our technology under control, and perhaps at the same time transfer some of our heavier industry and resource-gathering activities into space, allowing Earth to heal its wounds? The

Today. In an infinitesimal fraction of the planet's history, our own species is reshaping the planet's biosphere.

latter is a long-term visionary prospect, but now we are used to thinking in terms of geological time periods, not the puny half-decade periods that concern shortsighted contemporary politicians and industrial planners!

So much has been written about the environmental problems facing Earth that there is no point in rehashing the details here. Fortunately, the problems have risen high enough in public consciousness that each day's newspapers and TV newscasts describe problems. Promising solutions are few and far between because of a built-in difficulty in recognizing the need for action.

THE TROUBLE WITH RECOGNIZING CHANGE

This difficulty stems from the shortness of the time scales we usually perceive. I've concocted a rule of thumb: if a change in conditions is slow enough to last more than three generations, the middle generation will fail to recognize that the situation is dynamic, not static. The first generation may recognize the exciting onset of a new era, and the last generation may live through a collapse of the previous order, but the middle generation tends to think that what they experience is the way life has always been. They may not recognize the seeds of their own destruction being sown around them.

There are countless examples of this effect in human societies. The Roman Empire might be mentioned. The American automobile industry of the fifties and sixties is an example closer to home, and one that involves Earth's resources. Living in a euphoria of cheap fuel costs (gas is much cheaper in America than in most other countries), Americans built bigger and bigger cars with bigger and bigger gas-guzzling engines. The generation that grew up in the fifties accepted this as the normal order of things, and scoffed at the Volkswagens and Toyotas of other countries, while foreign manufacturers tooled up to build small, efficient cars. When the gas shortages of the 1970s hit, big cars were as competitive as dinosaurs in the Tertiary. American car companies reeled and wailed as buyers downsized to better-built smaller cars. But throughout this period, and even today, many Americans bemoaned the end of "normalcy." Detroit, cheered from the sidelines by the car magazines, keeps trying to reintroduce powerful gas-guzzlers. The realization has still scarcely dawned that America's cheap mid-century energy costs were a product not so much of national superiority as of the fact that the American economic system was perfectly adapted to exploit geologic frontiers. Namely, it was extremely efficient at discovering and extracting the most easily removable petroleum and ores, first from our continent, and later from continents around the globe. Whether this economic system is adapted to an era with no terrestrial frontiers and with finite terrestrial resources remains to be seen.

Some analysts suspect that the history of the American automotive industry is an allegorical history of the whole American economic system. For generations, from 1800 to 1950, American society was unique for having both (a) an exploitable frontier, as it swept across a virgin continent brimming with material resources, and (b) an efficient capitalist economic system fine-tuned to seek out and exploit those resources. Fortunes were built by finding and selling coal, oil, iron ore, copper ore, and other riches of the planet's crust. Americans prospered and gained a higher material living standard than other countries (by which Americans mean, in effect, a greater rate of resource consumption).

Generations of Americans, especially those from the 1940s to 1960s, came to view this standard of living as normal, a simple result of (a) the inborn superiority of the American system, or worse yet, (b) American moral superiority. They developed another peculiarly American myth—namely, that the reason for the success was wholly factor (b), not the combination of (a) and (b). They shied away from the fact that assuming moral superiority is often a fatal mistake.

Perhaps the most important lesson to be learned from the history of Earth is that our situation in life is not static, and that we (personally and as a species) have to be on the lookout for new environmental niches. The above rule of thumb applies also over geologic time scales. We are lulled into the misperception that our present world situation is normal, the way life has always been, instead of remembering that the Cenozoic Era represents only 1.4 percent of the planet's history. During the cosmic transitional instant of the last 4000 years, we have accumulated a cultural heritage based on the notion that this is the normal condition of Earth. Earth itself is a fixed stage in the background of our play. The drama of life goes on indefinitely, with humans cast in the role of lords over various subordinate species who have been around since the Garden of Eden was stocked with them.

Of course, a closer reading of Earth's history tells us a very different story.

In our lifetimes a new view is emerging as we recognize some disquieting facts. Our burning of Earth's fossil fuels has artificially increased the carbon dioxide content of the atmosphere; this in turn is probably changing the climate of our planet; there have been dramatic extinctions of species in the past; our planet had a different atmosphere in its early past; and as if this weren't enough, the planet Mars apparently went from an early history with liquid water to a later history of frozen, dry conditions. All of these discoveries tell us that planets change and that we should not view ourselves as inevitable, permanent fixtures.

WHAT IS "NATURE"?

There is a more subtle and philosophically troubling point. The environmental movement mushroomed when we saw our planet from space during the Apollo flights to the moon in 1969 through 1972. Seeing Earth as a little ball in empty space made people aware of the dangers of damaging our natural environment and of the need to preserve nature. But what is "nature"?

Virtually all the rhetoric of the environmental movement has implied that "nature" is some fixed reality provided by God, Earth itself, or cosmic forces. This "nature" is defined as somehow being an optimum state of Earth, in balance as long as it has not been tampered with by humans. This is the exact opposite of our conclusions above. There never has been a "fixed reality" for Earth. Continents move, mountains wash away, species come and go.

As if that weren't confusing enough, our very perceptions of what is pleasing in nature change from century to century; they are at least partly socially determined. Surprisingly enough, we *choose* much of what pleases us in Earth's nature, contrary to intuitive notions about some intrinsic standard of natural beauty. For example, as we saw in Chapter 1, the romantic revolution changed our attitudes about mountains and storms from feelings of foreboding and terror to emotions of wonder and exhilaration. Today's hikers are dismayed to find a road cutting through their favorite forest wilderness, but Thoreau extolled the virtue of country roads, always drawing the walker around the next bend. We recoil at the sight of strip-mined hillsides, but W. H. Auden wrote poems about the somber beauty of the blasted, strip-mined hills of his Cornwall countryside. Texas scholar Frederick Turner has pointed out other such examples in a remarkable essay (see the references). In his most remarkable passage, he asks us to imagine that the Grand Canyon, or something like it, was (as is plausible) the result of erosion caused by a mammoth strip-mining project. He describes how schoolchildren might be brought to the rim not to view nature's grandeur but to view in disgust this monstrosity of technology gone mad. The beauty is not just in the object, but in the mind-set.

So what sort of "nature" are we trying to preserve? Of course, in cases of modern urban sprawl and pollution, it seems obvious what nature we want to restore. Yet from a larger historical perspective, we realize that much of our environmentally noble thinking is skewed. The closer we inspect our concept of nature, the more likely it is to dissolve like a mirage. The beautiful, poplar-studded vistas of a southern European "natural" landscape are *not* natural; they are the product of thousands of years of clearing of the forest and farming. On the margins of urban-sprawling southwestern cities, environmentalists try to save the "natural desert," but this desert is not the same as it was when the Spanish conquistadors arrived. It has been altered throughout the past hundred years as rivers were destroyed by the lowering of the water table when groundwater was pumped, and many of its one-time grasslands were ravaged by overgrazing.

WHEN WAS AMERICA LAST "NATURAL"?

A dramatic challenge to our whole idea of nature in North America comes from recent archaeological work. Perhaps in the last paragraph, I conned you into accepting that North America was in its truly natural state when Columbus arrived. But even that is questionable. Fifteen thousand years ago—that is, 0.015 My or three ten-thousandths of a percent back in the history of Earth—North and South America were swarming with big mammals: the great, elephant-like mammoth and the forest-dwelling mastodon, a type of giant beaver, native horses, camels, certain now-extinct pronghorns and antelope, the ground sloth, and the saber-tooth cat. *That* America was the product of evolution working in tandem with continental drift—a reasonable definition of "natural." That may have been the last "natural" America, if we choose to think of nature as connoting Earth without the effects of humanity.

Curiously, 11,000 years ago in North America, about thirty-one different genera of large mammals became extinct within about 1000 years. This was just at the time when humans arrived across the then-existing land bridge between Asia and Alaska. Researchers used to assume the human hunters had nothing to do with the extinctions. After all, stone age hunters had pursued these game in Europe and Asia for thousands of years without driving them to extinction. It was therefore assumed that the sudden appearance of humans here could not be blamed for the extinctions. These must have

12,000 years ago. Colonization of North America. Glacial ice sheets tied up water and sea levels were lower, allowing Asiatic hunters to cross the land bridge into North America. Here they exploited the vast hunting grounds of the large mammals that had not yet learned to fear human predators.

resulted from some natural factor such as climatic changes associated with the retreat of the glaciers.

Since 1973, however, paleoecologist Paul S. Martin and others have reexamined this assumption. They stress that these mammals disappeared all across North America just as the main wave of human newcomers* arrived from Asia, 12,000 to 11,000 years ago. These early hunters constituted the first large immigration into North America. More important, they were the first with substantial technology for hunting big game, as is clear from their large, beautifully worked Clovis and Folsom stone spear points. A number of the most recent known mammoth remains have been found in association with Clovis and Folsom campsites, and these types of spear points are found across much of the continent. The large spear points stopped being made after the extinctions and gave way to smaller points for hunting smaller game; this indicates that the large points were made specifically for killing the large game. So we know that the first large wave of Americans were after big game.

Martin and others believe that the extinctions were caused not by nature, but by these hunters. The factor that was different in North America than in Europe, where mammoths and humans coexisted for many centuries, is that the American animals were completely unaccustomed to the new predators. As Martin wrote in 1973,

> The Paleolithic pioneers that crossed the Bering Bridge out of Asia . . . found a productive and unexploited ecosystem of over 10 million square miles. . . . It seems likely that, when entering a new and favorable habitat, any human population, whatever its economic base, would unavoidably explode. . . The hunters who conquered the frozen tundra . . . must have been delighted when they first detected milder climates as their route turned southward.

*It seems hardly appropriate to use the "politically correct" term, Native Americans, in this context; "Indians" is, of course, also a misnomer. These early Americans were Asiatic nomad hunters—the true discoverers of North America.

Today. Just as early humans explored outward from their point of origin and spread from one continent to another, we live at the moment of first exploration of our cosmic environment.

In Martin's theory, the migrant hunters swept forward, each generation moving farther into the new happy hunting grounds, decimating the big game, which had built up no defenses. The new, human population hunted its way forward at an average of roughly 16 to 25 kilometers per year (10 to 16 miles per year), spreading from Alaska across North and South America in only about a thousand years. Since they hunted primarily the mammoth and a few other animals, why did all the other large species die out as well? Martin's theory is that the whole ecosystem was upset by the destruction of the large prey, explaining why predators like the saber-tooth cat died out at the same time. Martin proposes a growth rate for the human population as high as 3.4 percent annually (the same as for the community who settled uninhabited Pitcairn Island after the mutiny on the *Bounty*) until they had spread all the way to South America.

If this view is correct, it revises our ideas of "nature" in America. It would mean that contrary to common American belief, Europeans never saw a "natural America," safely stewarded by the "native" American Indian population. Natural America, in the sense of an America untainted by human disruption, was an America with woolly mammoths that was demolished 11,000 years ago by early American immigrants.

Martin's view of the decimation of the big game species is not unlike what is happening on the African continent during the current century, as elephant and other populations decline due to hunting with modern weapons, ivory poaching, and similar factors.

The point is not whether mammoths were wiped out principally by humans, but rather that in many parts of the globe the "natural balance" between humanity and "nature" was disturbed much earlier than we usually assume. The image of "nature" that we developed during 3000 years of civilization is itself the product of Paleolithic hunters and Neolithic farmers who altered nature.

A UNIQUE MOMENT

All of which makes me believe that we must learn to be more conscious of the fact that humanity is not a steward of Earth's original, fixed "natural state," because no such thing existed. Rather (as the field of ecology tries to tell us), we are an active part of an active, changing nature. Which means that our job is not to stop civilization but to control it: to accept responsibility in making choices about what kind of civilization we build.

In my own view, such a civilization would use technology to optimize life and knowledge, while at the same time promoting simplicity in individual lifestyle instead of the current consumerism-induced complexity. Simplifying individual lifestyles seems to me a key ingredient in reducing the toll we are taking on planet Earth.

Technology was at a lower level millennia ago than today, but it acted over longer periods of time and caused greater changes than we usually admit. We have learned to accept those changes, which were relatively benign. The enemy of humanity is not so much technology itself, but rampant technology with uncontrolled accidental, unmonitored side effects. A special danger is the current indiscriminate freedom of large companies to manufacture and sell all sorts of products without much (if any) thought being given to any possible long-term damaging effects they may have. We have virtually no mechanisms for rationally considering the consequences of new chemicals and products. Scientists come along with their measurements after the fact, often too little and almost too late. As Julian Barnes says in his peculiar novel, *A History of the World in 10½ Chapters*, "That's what's wrong with the world. . . . We've given up having lookouts."

If the planet is changing all the time, why worry about the environmental changes we are causing? The answer has two parts. The first part is the more obvious: for the first time we are beginning to make environmental changes that are not lost in the noise of planetary evolution. Take carbon dioxide, for example. No longer do we light a mere campfire here and there. Our smokestacks are producing enough carbon dioxide to make a significant increase in the atmospheric content of that gas, and hence the greenhouse effect, yielding significant changes in the *global* climate. For a while we have a higher standard of living and a few people get rich, but in the long term, the planet and future generations suffer.

The second part of the answer is more subtle but also more important. It invokes time scales. We have caused these changes during civilization's fraction of a percent of Earth history—our hundredth of a second before midnight. The time scale for evolutionary change is hundreds of thousands or even millions of years. Environmental changes on a time scale of hundreds of years are therefore too fast for us; life can't adapt that fast by evolution. As terrestrial life found out during the Cretaceous-Tertiary transition, if things change too fast, you don't survive.

Our problem is not so much the fact that we are changing the environment, because that happens all the time through slow, natural processes. Our problem is that our civilization is changing it very fast, and worse yet, the change is out of control. We are living, therefore, in a unique moment in Earth's history.

THE NINE DANGERS

As remarked earlier, the major problems facing Earth in the near term are well publicized; we'll avoid a detailed examination here. However, a handy shopping list of nine dangers and suggested solutions may put things in perspective. These are the main threats that need to be overcome if our civilization is to navigate though the hundredth of a second after midnight.

1. Nuclear war. The threat of nuclear war is the lethal cliché of our age. With the advent of *glasnost*

Near future? Alteration of climate is already causing shifts in the biology of different areas. One problem is desertification, the expansion of mid-latitude deserts such as the Sahara, due to warming and shifts in rainfall patterns.

and the attempts of the superpowers to establish rapport, the threat seemed to have diminished. But this is a short-term view. The Iraqi war dramatized that around the world, smaller, less stable countries have acquired missiles and have approached or acquired the capability of building nuclear weapons. The danger to the planet as a whole is not in madmen launching a few nuclear weapons but rather in their unwittingly precipitating a massive unleashing of the superpowers' nuclear weapon stockpiles. Following pioneering work by Carl Sagan and others, scientists recognized that the major global result of such a war would be not just the explosions but the ensuing nuclear winter—the climatic cooling caused when explosion-ejected dust and smoke from countless fires block sunlight and darken the surface. Nuclear winter would make the 1816 year without summer look like a minor cold snap. The damage would be spread beyond military targets to the whole biosphere.

Miller

Researchers around the world now agree on these general consequences, if not the frightening unknown details. For example, a study published by a nongovernmental committee of Soviet physicists, biologists, and medical experts estimated that in an all-out global nuclear war, 25 percent of the population of North America would die, 37 percent in Europe, and 27 percent in Asia. Many more would sustain radiation lesions. Longer-lasting nuclear winter and radioactive debris would affect animal and plant species as well as humanity, endangering Earth's ecosphere in general.

Solutions. One part of the solution is to revise our notion of peace and peace movements. In our popular culture of bumper stickers and TV commentary, "peace" seems to be conceived of as a static condition, as the mere absence of war, achieved merely by defeating militarists who want war. After traveling in the U.S.S.R., Europe, Japan, and Australia, as well as in North and South America, I see peace as a much more active condition in which we must *engage* in positive programs of international commerce, science, and culture. Such programs work toward productive ends, and offer positive goals more profitable and interesting than going to war or trading in arms. Closing down the arms trade doesn't mean throwing thousands out of work. Why can't the superb engineering teams at Boeing and McDonnell Douglas be unleashed on designing solar power systems to be marketed in the United States and abroad? Why can't we promote more international manufacturing ventures, perhaps combining design teams from developed countries with manufacturing plants in less developed countries, aimed, for example, at developing the clean-energy-

If the average temperature of Earth continues to increase, it could result in the melting of the polar ice caps. The resultant rise in sea level would threaten low-lying coastal areas around the world, such as Miami Beach, whose glamorous and expensive hotels and condominiums would become uninhabited islands.

generating capacities of all countries? It takes only some international agreements, a modicum of political leadership, and an evolutionary shift of government contracts from giant military contracts to peace alternative contracts. Countries profiting from such ventures have little incentive to go to war.

2. Greenhouse effect and global warming. The greenhouse effect is caused by the fact that carbon dioxide and certain other gases in the air allow sunlight to pass through them to the soil but block the infrared heat radiation emitted by the soil back toward space. This means that the atmosphere absorbs the outgoing heat and warms up to a new equilibrium temperature that may be several degrees warmer than before. Our problem is that modern civilization is emitting more and more CO_2 as we burn fossil fuels, especially for manufacturing purposes, and the desire of the third world countries to have everything the developed countries have means that the problem will grow. All this is compounded by the destruction of forests (which consume CO_2 and produce oxygen), especially tropical rain forests.

During the 1980s, there was considerable debate as to whether the recent warm years marked the beginning of the predicted global warming. The issue isn't resolved, but new data suggest that the answer is, unfortunately, yes. In 1991, two separate research groups released the conclusion that 1990 was even warmer than the eighties, with the highest global average surface temperature in more than a century of weather measurements (*Science News*, *139*, p. 36). Based on the data of the two groups, six or seven of the warmest seven years in more than a hundred years of records have occurred since 1980.

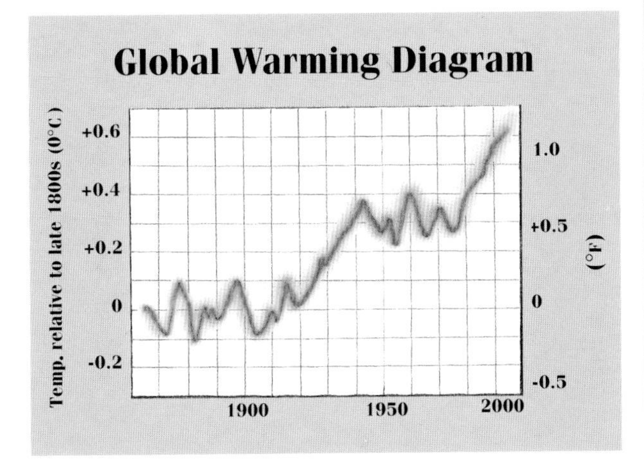

The CO_2 emissions and greenhouse effect were listed among the greatest environmental risks by a scientific board advising the U.S. Environmental Protection Agency about possible new programs in 1991. Interestingly, however, according to public polls at the same time, the public perceived global warming as far down on the list of ecodangers, below hazardous waste disposal and other problems. Furthermore, although many nations have called for CO_2 controls, and more than two dozen industrialized countries have pledged to limit their CO_2 emissions, the American government during the eighties and early nineties has downplayed the CO_2 threat. Only at a UN-sponsored meeting of a hundred nations in Virginia in early 1991 did U.S. officials finally agree that "despite large uncertainties the potential threat of climate change justifies taking action now," but the United States did not agree to controls on CO_2. (See the reports on climate politics by Kerr, 1991, and White, 1991, in the references.)

In a sense, global warming threatens world economic order more than it threatens the planet itself. Biological zones may simply move to higher latitudes, but countries currently favored may find themselves in trouble. If the CO_2 emission trend continues (which it surely will, at least for a generation), climatic warming will probably continue, in some places more than others. Deserts will expand in countries in the lower and mid-latitudes, including the United States, while high-latitude countries such as Canada and the U.S.S.R. may actually benefit with longer growing seasons. A side effect will be a sea-level trend opposite to that of the ice ages. Instead of seawater freezing into polar ice and thereby causing lower sea levels, we would have the polar ice melting, causing higher sea levels. Sea-level populated regions from New Orleans to

Bangladesh would be strongly affected by even a small rise in mean sea level. Although developed countries could cope with such changes, small, coastal countries of the third world would suffer disproportionately.

Not all scientists agree on the future trends. Climatologist Thomas Karl and his co-workers (1991) review various computer projections of climate trends, based on current growth of CO_2 emissions and other trends. Some of these models predict warming by 2° to 4°C (4° to 7°F), up to 15 percent increases in winter precipitation, and 5 to 10 percent decreases in summer rainfall in the central United States by the year 2030. On the other hand, Karl and some others conclude that random year-to-year changes are still greater than the changes due to greenhouse gas pollutants, and therefore that the year-to-year random effects will mask most of the greenhouse effects of climate change until after 2030. In any case, it seems that our world is headed for significant climate change as our children mature, with consequent changes in agricultural productivity, global economy, sea level, and social structure.

Solutions. Crash programs to develop energy sources alternative to fossil fuels, such as solar, wind,

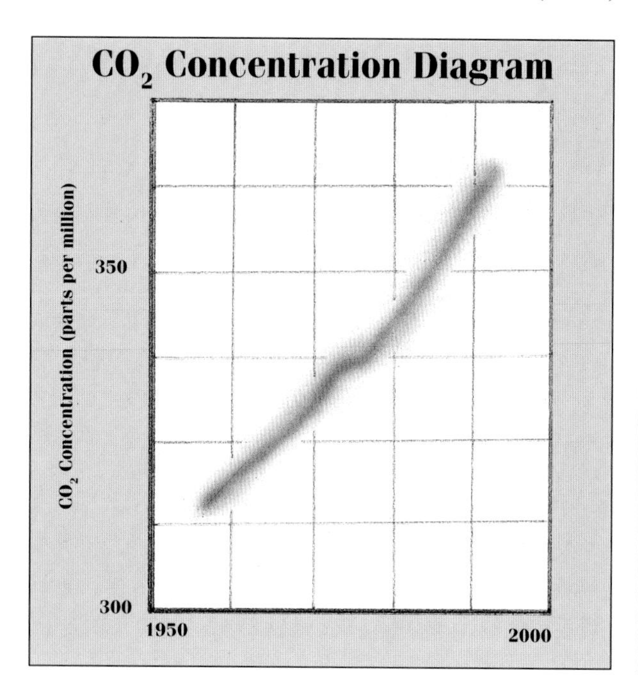

and geothermal power, perhaps through economic incentives from the U.S. Energy Department (which was allowed to languish in the 1980s) and similar agencies in other countries. Visionaries have proposed an international treaty to address this problem. It would calculate a maximum allowable global CO_2 production budget for the planet, and then divide this into transferable CO_2 production rights assigned to different countries based on their current production rates. Developed countries would find themselves assigned rates much lower than their current production, but undeveloped countries would have more than their current production. Advanced countries could then purchase CO_2 rights from third world countries, and the money could be used to set aside rain-forest preserves and develop alternate energy sources. We all lose a bit by being forced to change our economic lifestyles, but we all gain (or our grandchildren gain) by ensuring a more stable economic future.

3. Resource depletion. This is linked to the CO_2 greenhouse effect in that both problems involve our insatiable hunger to consume the wealth of our planet. Digging ever deeper, we've exhausted most of the easily accessible petroleum and mineral deposits in our search for energy and metals. One careful study of oil production in the United States, factoring in the rate of discovery of new oil fields, rates of pumping, and other factors, predicted a decline of 17 percent in production by 2000, as compared with production in the eighties. This projection was considered conservative, and the researcher stated that a more likely decline was 29 percent. We now accept lower-grade fuels and ores and dump the waste products of processing them into the biosphere, where they accumulate.

Solutions. Develop alternative energy sources, especially solar. Explore near-Earth space, especially the many asteroids that contain pure nickel-iron alloys and other metal resources. Develop solar energy farms in space, where twenty-four-hour-a-day sunlight is available, and process metals and other space resources there.

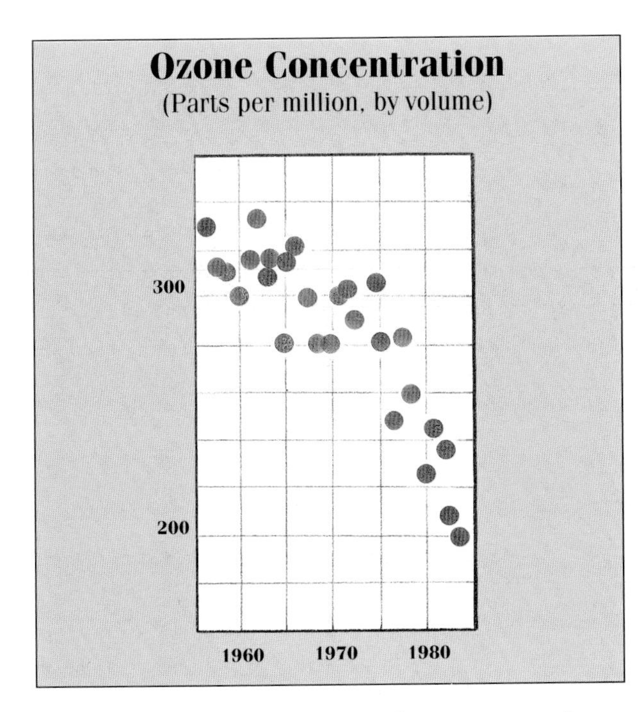

4. Damage to the ozone layer. Again, this is linked to the previous problem. Our hunger for consumer goods has led to the development of chlorofluorocarbon chemicals that are useful as refrigerants, aerosol propellants, and cleaning solvents, but when released into the atmosphere, these cause chemical reactions that break down ozone, a form of oxygen involving three oxygen atoms per molecule (O_3) instead of the normal two. As we have seen, ozone is crucial to blocking the sun's ultraviolet rays, which break down organic material and are harmful to life.

Solutions. Stop producing chlorofluorocarbons. This is already underway, with a 1978 ban in the United States on their use as aerosols, a 1985 UN-sponsored program, and a 1987 protective protocol signed in Montreal. However, the protocol leaves loopholes for developing countries to produce more of the damaging chemicals, which they want for their own economic development. An authoritative review by chemist F. S. Rowland, who was involved in discovering the damage to the ozone layer, says that the existing regulations will have "only a minor effect before the end of the century" (see references).

5. Hazardous wastes. We are already familiar with the acid rain that results from industrial waste chemicals emitted into the air in industrial areas. Related to the spewing of "junk chemicals" into the atmosphere is the production of additional waste by-products of heavy industry and power generation. Radioactive wastes from nuclear reactors are a classic example. Committees wrestling with plans to bury such wastes wonder how to keep the reposito-

Near future? Possible threats to civilization on Earth include comets and asteroids that pass close by and can collide with our planet. Collisions large enough to affect culture may happen every 10,000 years or so. Here a comet nucleus, a several-kilometer chunk of ice giving off jets of gas and dust, heads for Earth.

Near future? It is plausible (but not highly probable) that within a few centuries a major populated area, such as Europe, could take a celestial hit on the scale of the Siberian disaster in 1908. Here the outskirts of London undergo such a blast. Because heavily populated areas are only a small fraction of Earth's surface, such impacts in oceans or unpopulated areas are much more likely within the next few centuries.

ries sealed against earthquake, impact, vandalism, or human error during the centuries needed for the material to become harmless.

Solutions. Reducing our level of energy production and technology is one answer, but it suggests less flexible lifestyles for future humans. Switches to energy from solar, wind, and geothermal sources will help. Transferring the heavy industries of mining and smelting and some manufacturing into space, utilizing asteroidal materials and solar energy, may be long-term solutions.

6. Destruction of agricultural land through urban sprawl. Many major cities have been built along rivers and in sedimentary plains. The floodplain soils are among the richest agricultural soils, and as urban areas spread, these rich lands are being transformed to asphalted suburbias. It is ironic that the economic forces causing urban sprawl destroy Earth's most valuable lands.

Solutions. The trend toward urbanization may be reversed through technology. The proliferation of electronic communications and entertainment media may make it possible for more and more people to "return to the land," or at least to smaller towns in rural areas, while maintaining their traditionally urban jobs.

7. Population growth. All of the problems mentioned above are simply magnified as population densities grow. Earth has some carrying capacity for each species, dependent on the vitality of other species. Programs of chemical pest control, heralded in the 1960s as part of a green revolution that would expand agricultural production, have bottomed out, at least for now, as insects developed resistant strains and pesticides began to concentrate in farm animals, produce, and groundwater. It takes little imagination in an age of shrinking agricultural land and growing population to see unhappiness ahead. All the more reason to express surprise that some religions and political groups persist that not only discourage birth control but encourage having the maximum number of offspring. Of course, nature will not permit us to expand too far on Earth, and will take care of the problem for us through widespread famine if we don't take care of the problem ourselves.

Solutions. Better education; better understanding of voluntary birth-control practices and family planning to produce not just life but happy, productive life. In the very long term, creation of an interplanetary economy may permit new expansion of the human race beyond the confines of Earth.

8. Lack of respect for education. Another threat to the planet comes from the lack of respect for education in some sectors of global society. More is known about Earth, its history, the threats to its existence and their solutions than is considered interesting by the public and the media. Considering that the future of Earth and the future of civilization on it are in the hands of today's children, salaries of teachers are abysmal. Meanwhile, American televi-

Near future? A large meteorite streaks toward a populated area. An issue for strategic planners has been to distinguish modest-size meteorites from attacking missiles.

sion and other mass media glorify a macho culture of beer, gusto, brawn, and consumption instead of creation.

Solutions. We will need to vote, both in local elections and through our national representatives, to give more emphasis to education. Because educational TV programming does not pay its own way as easily as mass entertainment programming, we need to continue our support of public television and radio, despite repeated attempts to cut funding in these areas.

9. Asteroid impact hazard. All of the above threats to Earth arose from our own human activity. We have not included earthquakes and volcanoes on our list, because there is no firm evidence that any of them have been big enough to threaten the ecosphere in general. Asteroid impacts, however, are another story. The longer you wait, the larger an impact you are sure to get, as we've mentioned earlier. This is shown on the figure at right.

Another way of saying the same thing is that there are many different sizes of asteroids and comet nuclei out there in space, the larger ones being less numerous. NASA researcher Kevin Zahnle made a tongue-in-cheek table to rate the sizes of impacts we have talked about, and I have adapted it here.

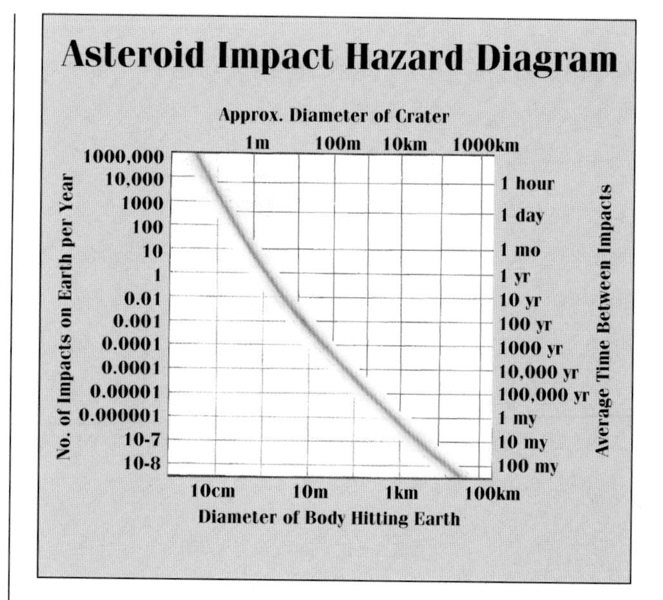

Every century or so, Earth gets hit by an object big enough to make headlines the world around. Every thousand years we get an impact big enough to cause a crater a few hundred meters across (if it hit a continent) and to do serious damage over a statewide area. It's as if we have a nuclear bomb (without radioactive fallout) exploding every few hundred years at random locations. Chapter 13 remarked that we can expect a 20-kilometer crater every 4 My, and a K-T-boundary-like impact big

ZAHNLE'S "FDA" SCALE FOR GRADING IMPACTS

Descriptive terms and lower-case descriptions are adapted from Kevin Zahnle; diameters of impactors and frequency of impact have been added by this author.

- SUPERCOLOSSAL (6000 km): once in history of Earth. Blows off enough material to form the moon.
- COLOSSAL (2000 km): only during first 400 My. Melts crusts; wipes out life.
- JUMBO (600 km): only during first 600 My. Vaporizes ocean; might wipe out life.
- EXTRA-LARGE (150 km): every 1000 My (?). Vaporizes surface biological zone; life might survive in oceans; forms largest multi-ring basins on moon.
- LARGE (60 km): every 500 My (?). Cauterizes continents; life forced back into oceans.
- MEDIUM (10 km): every 100 My. Kills dinosaurs; very dangerous.
- SMALL (50 m): every few centuries. Dangerous; front-page story; pray that it misses cities.
- TINY (boulder-sized pieces): every month. Makes bright fireball; can break through roofs.

enough to wipe out the majority of species every 100 My or 200 My. In other words, if we maintain an Earth-bound civilization and if we manage to get through the next few centuries without the kinds of eco-disasters mentioned above, we can be sure that an asteroidal impact will destroy civilization within 100 My or so!

Solutions. Develop the capability to operate in space, to discover such asteroids before they hit, and to change their orbits. This is not as impossible as it may seem, since the asteroids involved are small as asteroids go, and their orbits can be changed enough to miss Earth by relatively minor nudges if the nudging is done a long way away from Earth.

TWO PRIORITIES

Many of the solutions mentioned above touched on putting our engineers to work on two goals: (1) coming up with a range of minimally polluting alternative energy sources, and (2) developing the capability to explore options in space. The opportunities for space exploration and their effects on global society are beyond the scope of this book. Such aspects as space resources, asteroid mining, solar energy development, environmental effects, and the economic consequences of vigorous space reconnaissance were described in detail in our earlier book *Out of the Cradle*.

The point here is that just as life once spread from the oceans into the new hostile environment of the air and dry land, and just as humans spread from some ancestral home (Africa) across oceans and around the planet, so, too, we seem destined either to establish an interplanetary civilization or to die. What was once the speculative province of science fiction becomes more plausible as we look at the history of our planet in the long term.

1 to 6000 My from Now

17. The End of the Story

Some say the world
will end in fire,
Some say in ice.
From what I've tasted
of desire
I hold with those who
favor fire.

Robert Frost, "Fire and Ice," in the collection New Hampshire, *1924*

lanet Earth's end has been a matter of morbid curiosity for many generations. We can predict some important events in the planet's future. We can even predict that if Earth lasts long enough, the sun's final demise will first broil the planet and then turn it into a frozen waste. However, we cannot be absolutely sure that Earth's fate will be ruled by these predictable forces.

How do we know, for example, that the sun's collision with another star won't end Earth's saga first? The probability of collisions between stars can be calculated, and it turns out to be extremely low. We can assume that the sun will follow the normal course of cosmic evolution for a star its size. In that case, it will shine with slightly increasing light for another 5000 My, whereupon it will become unstable.

The question, then, is what will happen to Earth in that next 5000 My, and if Earth makes it through that period, what will happen as the sun ends its stellar career?

STILL MORE ASTEROID AND COMET IMPACTS

As emphasized in the last chapter, we can anticipate many more major impact events during the long haul of hundreds of My. As shown by the chart in Chapter 16, we can expect one or more impacts of the magnitude that caused the Cretaceous-Tertiary boundary extinctions during the next few hundred million years unless humans take control of the situation and modify asteroid and comet orbits to protect Earth.

To predict a little more about this situation, we can look at the sources of asteroid and comet impactors. A steady supply of modest-sized asteroids is kicked from the asteroid belt, between Mars and Jupiter, onto orbits that pass among the inner planets—Mars, Earth, and Venus. The process occurs primarily because of gravitational forces associated with the giant planet Jupiter. In a typical case, an asteroid perturbed out of the belt into the inner solar system may last 10 to 100 My on its new orbit before colliding with Mars, Earth, or Venus. As fast as asteroids of this group are swept up by one of these planets, new asteroids are kicked out of the belt to take their place.

To be kicked out of the asteroid belt by Jupiter, asteroids need to be following particular types of orbits in the belt. Generally, the largest asteroids, ranging from a few hundred up to 1000 kilometers in diameter, are on fairly stable orbits and are safe from being nudged toward Earth by Jupiter. Smaller asteroids, less than 100 kilometers across, are in plentiful supply in the belt, and many of them can be redirected toward Earth. At the present geologic moment, the largest Earth-approaching asteroid (cataloged as 1036 Ganymed) is 38 kilometers across, more than three times as big as the object that devastated Earth's species at the end of the Cretaceous Period. Probably we can expect even bigger potential impactors in future eons, ranging up to 100 kilometers across, big enough to devastate life on Earth.

But asteroids are not the only impact threat. Besides asteroids there are the ice-rich comets, which come from the outskirts of the solar system, beyond Pluto. The rate of comets hitting the Earth is believed to be similar to that of asteroids, which is why it is difficult to be sure whether a specific impactor, like the K-T object, was an asteroid or a comet. Most textbooks refer to comets as ranging from perhaps 1 to 20 kilometers in diameter, so they might seem to be less of a threat than asteroids. However, the textbooks are misleading. At least one remote comet (which does not come into the inner solar system near Earth) has been measured to have a diameter close to 40 kilometers. In 1988, astronomers D. J. Tholen, D. P. Cruikshank, and I discovered that a 200-kilometer object cataloged as "asteroid" 2060 Chiron, between Saturn and Uranus, had turned into a comet, blowing dust and gas off its surface! (The change in Chiron's behavior was due to the fact that it was moving closer to the sun and getting warmer. The weak heating changed some of Chiron's ice to gas, which expanded and blew off dust.) These observations prove that while small comets are more numerous than big ones, some big ones are still lurking in the outer solar system. Probably they come in a range of sizes, with a distribution of diameters similar to that of asteroids: many small examples and a few large examples.

The upshot of our discussion of asteroid and comet impact threats is that while Earth is unlikely to be hit by the largest asteroids, it might one day be hit by a very big comet. Comets are sent on Earthward courses entirely at random. Within some millions of years, comets much bigger than Earth-approaching asteroids may come through the inner solar system. For example, 200-kilometer Chiron comes close to Saturn every few thousand years and eventually could get deflected onto an Earth-approaching orbit. If Chiron doesn't get sent our way, there are more objects where it came from.

It seems likely that during future geologic time, and probably within 1000 My, Earth will be hit by both asteroids and comets much larger than the one that wiped out many species only 65 My ago. If so, Earthbound life would be devastated and evolution would literally pick up the pieces, starting over again from the molecular fragments of life. Probably other such events happened 1000 My ago or more, before the fossil record was complete enough for us to recognize their consequences.

THE OUTLOOK FOR "HUMANS": FUTURE EVOLUTION

Does this sound like a pessimistic reading of the future? We can report only what our best calculations show! Apparently Einstein was wrong: God does play dice, at least when it comes to slinging asteroids and comets toward the living Earth.

But if you are imagining cities full of our human descendants being demolished by a heartless cosmic game of dice 100 My in the future, think again. Recognizable humans have been around only 1 My or so. Monkeys and apes have been around for about 35 My, as we saw in Chapter 14, and tree-dwelling primates for about 50 My.

These figures raise the question: How long will recognizably human creatures be around? If we assume that evolution will continue as it has in the past, then we can confidently say that "humans" will be around for only 1 to 10 My, and that our descendants might better be regarded as post-human species, perhaps with greater intelligence (or with whatever other traits promote survival in the world we leave them).

On the other hand, it is plausible that evolution, especially human evolution, will be either retarded or speeded up by humanity itself. For example, virtually within our grasp is the possibility of the selection of genes in our offspring—rejecting those genes that cause horrific diseases and perhaps selecting for those that favor such aspects of intelligence as memory, spatial perception, or speed of reasoning. We will have to decide if we want to twist evolution in such ways. If such practices begin, then depending on the choices made, humans might either evolve more rapidly to unrecognizably intelligent states or perhaps stagnate in conscious or unconscious attempts to retain familiar "human" qualities.

Aside from questioning the biological makeup of our descendants 1 My or 10 My into the future, we can be sure that the technological basis of their civilization—if it exists at all—will be dramatically different from ours. After all, look at the shift from flint points to carbon chips in only 10,000 years—a mere instant of geologic time!

In the context of the incredible time scales of the *planet's* future, the most radical science-fiction concepts become mundane. In particular, two scenarios seem the most likely, but only one of them can happen. In the first scenario, we will demolish human civilization and life on Earth within a few generations, either through warfare or a stupid miscalculation that leads to a fatal environmental modification.

In the second scenario, we will establish an interplanetary culture, drawing on energy and material resources in space, and our civilization will ultimately spread among orbiting space cities and settlements on other planets. This would give humanity insurance against catastrophes on individual planets. Conceivably, to follow a science-fiction tradition, some of these planets and cities could be far outside our own solar system. The plot of Isaac Asimov's *Foundation* series of novels does not seem fantastic: humanity has spread across many star systems, and archaeologists are trying to identify which forgotten planet was the one that spawned our race.

As pointed out earlier, our own destiny is not necessarily tied to our planet's. Space enthusiasts, science-fiction writers, and visionaries emphasize that if we but learn to survive in human-made environments off Earth, we can survive any man-made or natural disasters that shatter the present ecosystem in the same way the Cretaceous or Permian ecosystems were shattered. That is why the visionary "Biosphere II" project, to build the first completely enclosed, self-contained human, animal, and plant habitat near Tucson, Arizona—a sort of desert

is the reason the sun was cooler in the past.

Calculations indicate that this process will continue for 5000 My. During that time, the sun may grow about two or three times as bright as it is now. Of course, this could warm Earth's climate considerably in the unimaginably distant future. Try to picture a time, say, 3000 My from now, when the sun has become twice as luminous as today. Barring effects of clouds and climate, the increased sunlight alone would cause the mean temperature of a now-temperate zone to rise from around 20°C (68°F) to a stifling 75°C (167°F).

However, most scientists believe Earth's climatic reactions are not so simple. Earth has a built-in buffer against wild climatic oscillations due to changes in sunlight. For example, as the sunlight increases, more water will evaporate from the oceans into the air. This means that more water vapor will be available to form clouds. As the cloud cover is increased, more sunlight reflects back into space from the bright cloud surfaces. The mean reflectivity of Earth, currently, averages about 39 percent—a combination of the reflectivity of sea, land, and highly reflective bright clouds. If Earth was entirely shaded by clouds, the reflectivity might rise as high as 80 percent. In our example, when the sun's radiation doubles, if the increasing cloud cover pushed the mean reflectivity only as high as 69 percent, this would entirely offset the increased sunlight and the surface temperature would stay about the same!

The same effect may have worked in reverse to buffer the effects of lower solar radiation in the past: less sunlight would evaporate less moisture, causing fewer clouds and more chance for the wan light to warm the ground.

Plants could also get involved. A mild increase in sunlight would make plants prosper, consuming more CO_2 and acting to reduce the greenhouse effect, thereby canceling the warming effect.

This touches on the famous Gaia hypothesis, named for the ancient Greek goddess of the Earth. This idea was introduced by British chemist James Lovelock in 1972. It is championed by several scientists, but it's highly controversial. In its mildest form, the Gaia hypothesis points out that climate effects are buffered not only by cloud formation, oceanic heat absorption, and other geographic processes but also by complex biological processes, and that the net effect is to keep Earth maintained at clement conditions for living things.

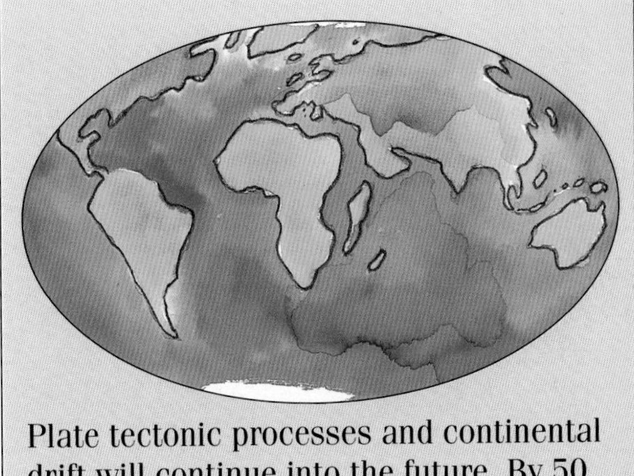

Plate tectonic processes and continental drift will continue into the future. By 50 My from now, one likely change will be the breaking off of eastern Africa along the modern-day rift valley. Similarly, Baja California will probably have become detached from the Mexican mainland.

The idea is that the biosphere acts like a single, giant, complex system to regulate itself. We already saw in Chapter 9 how life could evolve only in an oxygen-poor atmosphere, but that life caused the atmosphere to change toward more oxygen, which in turn fostered more advanced life.

The more mystical and controversial form of the Gaia hypothesis, sometimes championed by pseudoscience writers, portrays Earth as a single giant organism, with all its lifeforms linked to create an environment that the lifeforms enjoy. Lovelock's phraseology stakes out a middle ground: Earth's biosphere forms "a complex system which can be seen as a single organism and which has the capacity to keep our planet a fit place for life."

Among the criticisms of this hypothesis are

4000 My from now. The Milky Way and Andromeda galaxies have collided and passed through one another, like a pair of ghosts colliding at night. The sun, wrenched from its usual place by gravitational forces, spins off into the universe with its retinue of planets. Looking back into the night sky from near the ruins of some forgotten monument, we can see the entangled galaxies (the Milky Way nearest) receding as time-ravaged Earth follows the sun into intergalactic space. Mankind may have vanished from Earth by this time, as far into the future as life's origin is behind us. Perhaps humans finally rendered their home planet uninhabitable and the race died here, as the dinosaurs did; or perhaps mankind evolved into something very different. Or perhaps we have left Earth behind.

charges that it (1) pictures species evolving not in response to changes in climate but rather by looking ahead to see what conditions they want to create; (2) pictures species sacrificing themselves for the stability of the ecosphere, instead of competing in Darwinian fashion; (3) has not been put forth in a quantitative manner where the responsive effects can be calculated; and (4) is not easily testable by scientific observations (Mann, 1991).

The last charge is interesting philosophically. One quality of a "useful" scientific theory is that tests can be dreamed up to see if it is false. Usually a theory cannot be proven true by one test with a positive result, but it can be proven false by one test that disagrees with its predictions. The people who say the Gaia hypothesis can't be tested point out that when it comes to inhabited planets, we are dealing with the statistics of one, and we can't very well alter the temperature or other conditions to see whether the ecosphere self-corrects. But perhaps this criticism is too conservative. If astronomers detect planets around other stars, and if we ultimately discover inhabited planets, we may get a better idea of whether all planets' ecosystems evolve to maintain clement conditions, as the Gaia hypothesis predicts.

A more extreme view of Earth/Gaia as a mystic terrestrial life force might predict that only Earth, among all potentially inhabited planets, behaves this way. This too could be tested if we find many other populated planets.

Regardless of our view of Gaia, we know that not all feedback loops act to keep the climate mild. There could be destabilizing feedback processes—if not for Earth, then for some other planet. For example, if increased heat beyond some point killed a fraction of the plants, then less CO_2 would be consumed. Then the greenhouse effect would be aggravated and the heat would increase even more, and so on until most or all plants died. Most scientists still regard the Gaia hypothesis as unestablished, but they admit that complex feedback loops between biology and geology, which act in a way to influence climate conditions, are important in understanding the nature of our habitable planet. These feedback processes might be called Gaia effects. Regardless of how efficient the Gaia effects are in nature, it is clear that if we change the environment too fast on our own—for example, by increasing the CO_2 content too fast—the Gaia feedback effects won't be fast enough to mitigate the consequences.

A COLLISION WITH THE ANDROMEDA GALAXY—4000 MY FROM NOW

Our galaxy, the Milky Way, is a system of gas, dust, and 200,000,000,000 stars arranged in the shape of a flat disk. Only about 20 disk-diameters away is another similar disk of stars, the Andromeda galaxy. We know, from studies of other, more distant galaxies, that galaxies sometimes collide. Canadian astronomer Marshall McCall, mentioned above, has calculated the future paths of the Milky Way and Andromeda galaxies, and he believes they will collide in about 4000 My.

Just as two swarms of gnats might pass through each other, the two galaxies of stars will pass through each other without individual stars colliding. Individual stars are too small and far apart to have much chance to collide with each other during such an event. However, we will not get off scot-free. Intergalactic collisions are well known, and studies show that they strip the gas and dust from the two colliding galaxies. Furthermore, the gravitational attractions of the galaxies distort the the galaxy shape, ejecting many stars into intergalactic space. It is probable that when Andromeda comes a-calling, the sun will find itself, along with other stars, pulled well outside the plane of the Milky Way, perhaps even being torn permanently away from its parent galaxy.

Such events would radically change the night sky. For several hundred My before the collision, Earth creatures would see the Andromeda galaxy (which today looks like a pale fuzzy patch the size of the moon) filling the night sky with a faint pattern of spiral arms and a dazzling bright center, as bright as Venus. After the collision, if the sun was one of the stars torn out of the Milky Way, Earth creatures might see both galaxies filling the sky, and the center of one or both galaxies might be transformed into a brilliant quasar.

Strangely enough, Earth creatures might not notice other radical changes, because the whole solar system would respond to gravitational forces as a unit. Thus, if the sun was one of the stars torn out of the galaxy, the other planets would go along with it, circling the sun in their orbits as they always have.

THE FATE OF THE MOON

We have noted in several earlier chapters that the moon has been moving away from our planet, and that because of gravitational links between Earth and moon, this causes the length of the day to increase.

Will this process continue indefinitely, with the moon eventually drifting away into orbit around the sun? Probably not. As the moon drifts outward, the length of the month increases, but the length of the day increases even faster, so that the day would eventually match the month if the process went on long enough. That would happen with the moon at a distance of roughly 550,000 kilometers, about 1.4 times as far as it is now. The length of the day and of the month would be 47 "days" (i.e., 47 x 24 hours), and the moon would hang permanently over one continent, never to be seen from the Earth's other side. In this configuraton, tidal gravitational forces between Earth and moon would no longer be effective, and the moon would stop drifting out.

However, a weaker tidal force from the sun would still be slowing Earth's rotation. With Earth then turning slower than the moon moves around it, and with the moon rising in the west and setting in the east, the angular momentum exchange between the two bodies would be reversed, and the moon would start drifting back inward, toward Earth. Given long enough, the moon might eventually crash onto Earth.

The trouble with this scenario is that with the day lengthening by half an hour every 100 My, or five hours every 1000 My, it would take many thousands of millions of years for the moon to reach the point where its outward movement stops.

Before that time comes, the sun will have run out of fuel.

A GIANT SUN—5000 MY FROM NOW

It is also true that no matter how efficient Gaia effects are, they won't be able to deal with the final stages of the sun's evolution. As the zone of hydrogen "burning" inside the sun spreads from the hydrogen-exhausted core outward toward the surface, it will heat the surface layers, causing them to expand dramatically. So although the sun will have been growing brighter all along, the process will accelerate. In its final millions of years, the sun will get much bigger and brighter. This process will be aided by events in the sun's central core. Lacking any energy production of its own, the core will have inadequate heat and pressure to resist gravity; it will start to shrink and get hotter. Finally, it will get hot enough to ignite nuclear reactions, not among hydrogen atoms but among helium and even heavier atoms. The surge of heat from these reactions will cause the outer layers of the sun to expand even more.

The net result will be that, about 5000 My from now, the sun will turn into a huge, very luminous type of star called a *red giant*.

We know this is an accurate prediction because astronomers have observed hundreds of red giant stars as well as stars in transitional stages. The term *red* comes from the fact that the outer layers of a red giant are cooler than the sun's surface, and hence somewhat redder in color. They are red-hot instead of white-hot. Theoretical calculations give a detailed picture of the internal nuclear reactions and how they drive the star from the sun-like stage to the red giant stage.

The properties of red giant stars have been measured directly by telescopic observations. Red giants typically have a diameter 100 times that of the sun and a luminosity several hundred times that of the sun. Their outer "surface" layers are very thin gas, hardly more than glorified interplanetary gas. After all, the "surface" of any star, whether red giant or ordinary sun, is not solid like that of a planet but merely the layer of gas that is ionized and therefore opaque. The surface of a star is no more a surface than the "edge" of a candle flame through which you can move your finger.

THE DESTINY OF EARTH

The ominous thing about these numbers is that a star 100 times the size of the sun has a diameter almost as big as the orbit of Venus! Some red giants are known with orbits bigger than that of Earth; some are as big as the orbit of Mars! This means that the flames of the sun will one day reach out to lick at the Earth/moon system. If the sun grows big enough, it will ultimately engulf our planet. Earth would find itself inside the luminous, tenuous outer atmosphere of the sun.

5000 My from now. The sun will eventually expand and turn into a red giant star. As this happens, the surface of Earth will be heated and lava flows will convert much of the surface to magma ponds, if not magma oceans.

6000 My from now. If Earth is not engulfed in the expanding red giant sun, it will witness the red giant running out of stellar fuel. When that happens, the sun will collapse down to a star not much bigger than Earth itself—a so-called white dwarf. The luminosity of the white dwarf is so little that Earth's surface will be entirely frozen.

Several things will happen to Earth before that happens. As the sun becomes more luminous, the Gaia feedback effect of ocean water and cloud cover will cease, because the ocean water will eventually all evaporate. Earth will have regressed back through its infant stages, with higher temperatures and a steamy, murky atmosphere.

Eventually, increasing heat will drive the atmosphere off into space. The less air and cloud, the lower the mean reflectivity of Earth and the hotter the surface, because rock reflects less radiation than the cloudy atmosphere—a feedback that will exaggerate the effects of the sun's warming instead of canceling it. Finally, bare rock will be exposed to the blinding sun. As the sun becomes a red giant with luminosity 300 or 400 times the present value, the mean temperature of Earth's surface will reach that of molten lava. Rocks and soil will melt. Mountains will turn to molasses, sag, and level out.

Now airless, Earth will once again have acquired

a magma ocean. The surface will be an endless flat sea of molten rock, dominated by a vast red sun filling much of the sky.

Interestingly enough, while Earth is rendered sterile and molten during this stage, the icy planets of the solar system will go through clement periods for perhaps hundreds of My. For example, Saturn's moon Titan, with a thick nitrogen atmosphere and a present-day temperature of –180°C (–292°F), could warm to room temperature and even beyond, to the temperature of boiling water. Even Pluto could rise to room temperature.

If the sun continues to expand to a diameter as large as Earth's orbit, our now-molten-surfaced planet would find itself skimming through luminous, thin gas—the sun's tenuous outer layers.

If the sun expands enough and for a long enough time, Earth would experience the same kind of drag effects as a satellite skimming the atmosphere. It would spiral inward. Some calculations suggest that it will ultimately spiral all the way into the sun's interior. If so, the atoms of our bones, which will have long since merged with the atoms of Mother Earth's soil and atmosphere, will finally merge with the atoms of Earth's father, the sun. That would be the end of Earth as a distinct entity.

A DWARF SUN—6000 MY FROM NOW

If the sun doesn't expand enough to cause too much drag, a different fate would await Earth. The red giant stage of any star lasts only a limited time. Finally, as the sun runs out of fuel, it will become unstable. No longer will it be producing heat and radiant energy that lead to an outward force, balancing the inward force of gravity. Gravity will cause it to shrink down to a very small size, about the size of Earth itself. An exhausted star of this type is called a *white dwarf*, "white" referring to the fact that the tiny star has about the same color as the sun.

Like red giants, white dwarfs have been carefully observed and are well understood observationally. A typical white dwarf has only one-thousandth the luminosity of the sun and slowly fades during thousands of My to one ten-thousandth the luminosity of the sun and even fainter.

If Earth escapes spiraling into the sun, it would end its days as a cold, dimly lit ball of rock. With the sun reduced to a white dwarf at one-thousandth of its present luminosity, Earth would have a surface temperature around –217°C (–359°F). Any gases that escape from a still-warm interior would condense into frost deposits on the surface, or form a thin, frigid, sterile atmosphere.

The galaxy itself will spin on for many thousands of My into the future, long after the sun and solar system have run their course. In other parts of our galaxy or in other galaxies, new suns will still be forming, just as they are today. Other Earth-like planets may be flourishing around some of them. The destiny of terrestrial life in such a distant future is almost beyond imagining. Will it have evolved beyond recognition? Probably. Will it have found new homes on planets near other stars? Possibly. We don't know. Such life might regard ancestral Earth with no more awe than we reserve for Africa, our probable ancestral continent.

From studying Earth, we have learned that continents and species are transient. Now our planet and our sun teach us that Earth as we know it is transient, in the long run of time. Is this depressing news? Not if, as Native Americans would say, our hearts are right. As individuals and as a species, our job is to make life as transcendent as we can. Meanwhile, as our planet evolves and changes, some Earth-like planet in an unfamiliar star-swarm of some galaxy will witness strangely familiar scenes: struggling creatures, having just evolved intelligence and self-consciousness, will be living out lives of happiness and hope, sorrow and wonder—lives not altogether unfamiliar to us.

GLOSSARY

accretion: aggregation of planetary bodies during collisions of planetesimals.

asteroid: an interplanetary body of rocky or metallic composition.

Archaeozoic Era: a designation sometimes used for the era of earliest development of life, prior to development of prominent fossil-leaving lifeforms; roughly 4000 to 670 or 550 My ago. Also called, loosely, the Precambrian Era.

basalt: a dark gray or brown lava rock that forms much of the Earth's oceanic crust; one of the most common and important types of igneous rock on several planets.

C

caldera: a large volcanic crater, usually formed by collapse and larger than a few hundred meters across.

Cambrian Period: the geologic period from about 550 to about 500 My ago, traditionally known for the earliest complex fossils, including especially those of trilobites. (But see also EDIACARIAN PERIOD.)

Carboniferous Period: the geologic period from about 360 to about 290 My ago, known for dense land forests that left massive coal deposits in some regions.

Cenozoic Era: the most recent of the major eras of geologic time, from 65 My ago to the present.

coacervates: nonliving cell-sized clumps of organic material; possibly a step toward evolution of life.

comet: an interplanetary body containing a large amount of ice as well as sooty and rocky material.

continent: a large mass of relatively low-density and granitic rock, floating on the mantle.

continental drift: the theory that continents move relative to each other; expanded and replaced by the more general theory of plate tectonics.

core: the nickel-iron central region of Earth or other planets.

crater: a circular depression with a raised rim, caused by either meteorite impact or volcanism.

Cretaceous Period: the geologic period from about 150 My ago to about 65 My ago; the last period in the Mesozoic Era, climaxing dinosaur evolution and ended by the K-T impact.

Cretaceous-Tertiary boundary: See K-T BOUNDARY.

Cryptozoic Era: a designation sometimes used interchangeably with ARCHAEOZOIC for the era of evolution of earliest life, before good fossils were formed, from roughly 4000 or 4500 My ago until about 670 or 550 My ago.

crust: the thin (several kilometers to several tens of kilometers) layer of low-density rock at the surface of a planet.

D

Devonian Period: the geologic period from about 405 to 360 My ago, marked by abundant marine life and sometimes known as the age of fishes, but also marked by proliferation of some early substantial land plants.

DNA: deoxyribonucleic acid, a giant molecule in the nucleus of the cell, fundamental to cell reproduction and carrying the genetic information that controls structure of off-spring cells.

Ediacarian Period: the geologic period from about 670 to about 550 My ago; recognized in the 1970s and 1980s from deposits in the Ediacaria Hills of Australia, with some of the first fossils of complex organisms.

era: a broad division of geologic time, subdivided into periods.

eukaryote: a type of cell with a well-defined nucleus; evolved possibly about 2500 My ago from earlier prokaryote cells.

fault: a fracture in Earth's crust along which there has been relative motion of the two sides of the fracture (and along which there may be further future motion). Typically, earthquakes occur along faults.

fundamentalist: a person or a philosophic tradition that tries to derive knowledge about nature, not by direct observation of nature but by assuming that certain ancient writings are literally true.

Gaia hypothesis: a controversial idea, discussed in both extreme and mild forms, that Earth acts like a single symbiotic living organism and adjusts its ecosphere conditions through various feedback effects to optimize conditions for life on its surface.

geologic time scale: the succession of geologic periods, with their absolute ages, throughout Earth's history; sometimes used interchangeably with STRATIGRAPHIC COLUMN.

Gondwanaland: a loosely defined giant continent in the Southern Hemisphere about 450 My ago, combining South America, Africa, Antarctica, Australia, and India; named for an Indian province.

giant impact theory: the theory that the moon formed as the result of impact between Earth and a very large planetesimal, blowing mantle material into orbit to form a satellite.

granite: a moderate-density igneous rock, rich in silica, composing much of the continental crust of Earth.

Great Dying: the strongest known wave of extinctions, at the end of the Permian Period 250 My ago, when 90 percent of all existing species died out. Cause is unknown.

greenhouse effect: warming of a planet's atmosphere by blocking of outgoing infrared "heat radiation," due to carbon dioxide or certain other gases that absorb infrared radiation.

I

igneous rock: rock formed by solidification of molten magma.

isotope: a form of an element distinguished by the number of neutrons in the nucleus. Each element has several isotopes, some stable and some unstable.

J

Jurassic Period: the geologic period from about 210 to 150 My ago. Dinosaur heyday.

K-T: a traditional geological abbreviation for the boundary between the Cretaceous and Tertiary Periods, 65 My ago.

K-T boundary: the boundary layer between the Cretaceous and Tertiary periods, widely marked by a thin (inch-scale) soil layer containing an excess of iridium and other signs of a large impact.

K-T extinctions: the second-greatest known wave of extinctions, 65 My ago. Evidence indicates that it was caused by a large impact. See K-T IMPACT.

K-T impact: the impact of one or more large asteroids or comets that ended the Cretaceous Period and is believed to have wiped out the dinosaurs and many other species.

Laurasia: a loosely defined giant continent of the Northern Hemisphere about 450 My ago, combining North America, Europe, and Asia.

lava: magma on or near the surface of a planet.

magma: molten rock, especially when still underground.

magma ocean: a surface layer of molten material believed to have existed during planet formation, due to heating by meteorite impacts. It was first recognized on the moon, but is also believed to have existed on Earth and other planets.

mantle: the large, interior region of Earth and other planets lying between the metal core and the crust and composed of high-density rock.

Mesozoic Era: the geologic era of "middle life," from about 250 to about 65 My ago.

metamorphic rock: rock that began as igneous or sedimentary rock, but was altered in chemistry and/or structure by heat, pressure, or invasion of hot, mineral-bearing, underground water.

meteorite: an interplanetary body that falls through

the atmosphere and reaches the ground; usually restricted to bodies smaller than house-size.

Mid-Atlantic Ridge: a volcanic ridge down the middle of the Atlantic sea floor, marking the edge of two tectonic plates. America and Europe move apart as new lava is erupted along the ridge.

Neptunism: a now discredited theory from about 1800, that all rocks derive from deposits of materials formed in oceans.

Ordovician Period: the geologic period from about 500 to about 440 My ago, marked by early evolution of fishes.

P

Paleozoic Era: the geologic era of early life, from about 670 or 550 My ago (depending on usage) until about 250 My ago.

Pangaea: a term once used loosely to define a hypothetical primeval supercontinent, but now used for a giant continent formed by collision and partial merger of Gondwanaland and Laurasia about 250 My ago.

period: a subdivision of geologic time. Several periods make up an era.

Permian Period: the geologic period from about 290 to about 250 My ago; the last period in the Paleozoic Era. The Permian ended with the Great Dying, when 90 percent of existing species became extinct.

plate tectonics: tectonic movements of large, intact regions of Earth's crust, called plates; formerly known as continental drift. Movement is generated by sluggish currents in the mantle.

planetesimals: primordial rocky, icy bodies of the solar nebula, before the planets formed. The term usually refers to sizes from dust particles up to a few hundred kilometers in diameter. Asteroids and comets are leftover planetesimals, or their fragments.

Plutonism: a now discredited theory, from around 1800, that rocks formed primarily by crystallization of molten material produced by deep-seated volcanic heat sources.

Precambrian: a general designation for the time prior to the evolution of lifeforms leaving prominent fossils, roughly 4500 or 4000 My to about 670 or 550 My ago; somewhat loosely defined due to changes in fossil discoveries.

primitive atmosphere: a planet's initial atmosphere.

prokaryote: a type of cell lacking a well-defined nucleus. typified by bacteria and algae; probably the earliest type of cell, evolving perhaps 4000 or 3900 My ago.

radioactivity: the decay of unstable, or radioactive, isotopes.

radiogenic: caused by or related to decay of radioactive isotopes.

radioisotopic: utilizing radioactive isotopes, as in radioisotopic age determination.

RNA: ribonucleic acid, a giant molecule in cell nuclei; possibly an evolutionary precursor to DNA molecules.

S

secondary atmosphere: an atmosphere produced by release of internal gases of a planet, usually during volcanism.

sedimentary rock: a rock type formed by accumulation and cementing of particles, such as sandstone from sand grains or limestone from calcium-rich deposits on the sea floor.

Silurian Period: the geologic period from about 440 to about 405 My ago, marked by marine life and earliest substantial land plants.

strata: layers of rock formed by deposition of seafloor sediments, wind-blown sediments, lava flows, etc. (Singular: stratum).

stratigraphic column: a cross section (or imaginary drill core) showing the vertical succession of strata laid down at different periods of geologic history, with oldest on the bottom and youngest on top. (See GEOLOGIC TIME SCALE.)

stromatolite: algae colonies with a cabbage-like structure that left some of the earliest fossils, as old as roughly 3000 My and still found on Earth today.

tectonic: caused by or related to the movement of large masses of a planet's crust.

tidal force: a "stretching" associated with the difference between one astronomical body's gravity acting strongly on the near side of a nearby "target" body and more weakly on its far side.

Triassic Period: the geologic period from about 250 My ago to about 210 My ago, known for the rise of dinosaurs.

volcanism: eruption of magma at the surface of a planet.

volcano: a vent or fissure where magma is erupting at the surface of a planet.

REFERENCES AND SUGGESTED FURTHER READING

Alvarez, Luis W. (1987) "Mass Extinctions Caused by Large Bolide Impacts." *Physics Today*, July, 24.

Badash, Lawrence (1989) "The Age-of-the-Earth Debate." *Scientific American, 261*, August, 90.

Bakker, Robert T. (1987) "The Return of the Dancing Dinosaurs," in *Dinosaurs Past and Present*, eds. Sylvia J. Czerkas and Everett C. Olson. Seattle: University of Washington Press.

Beatty, J. K. (1991) "Killer Crater in the Yucatán?" *Sky and Telescope, 82*, 38.

Benz, W., A. G. W. Cameron, and H. J. Melosh (1989) "The Origin of the Moon and the Single-Impact Hypotheses III." *Icarus*, 81, 113.

Berggren, W., and J. van Couvering (1984) *Catastrophes and Earth History*. Princeton: Princeton University Press.

Brush, Stephen G. (1989) "The Age of the Earth in the Twentieth Century." *Journ. History of the Earth Society, 8*, 170.

Budyko, M., G. Golitsyn, and Y. Izrael (1988) *Global Climatic Catastrophes*. Berlin: Springer-Verlag.

Budyko, M., A. Ronov, and A. Yanshin (1985) *History of the Earth's Atmosphere*. Berlin: Springer-Verlag.

Cloud, Preston (1988) *Oasis in Space: Earth History from the Beginning*. New York: W. W. Norton.

Covey, Curt (1984) "The Earth's Orbit and the Ice Ages." *Scientific American, 250*, February, 58.

Davies, G. L. (1969) *The Earth in Decay*. New York: American Elsevier.

Dawkins, Richard (1976) *The Selfish Gene*. New York: Oxford University Press.

Drake, Michael J. (1989) "Geochemical Constraints on the Early Thermal History of the Earth." *Zeitschrift Naturforsch, 44a*, 883.

Editors, Encyclopaedia Britannica (1978) *Disaster! When Nature Strikes Back*. New York: Bantam Books.

Edmond, J. M., and Karen Von Damm (1983) "Hot Springs on the Ocean Floor." *Scientific American, 248*, April, 78.

Golitsyn, G., and A. Ginsburg (1985) "Natural Analogs of a Nuclear Catastrophe," in *The Night After. . . : Climatic and Biological Consequences of a Nuclear War*, ed. Y. Velikhov. Moscow: Mir Publishers.

Hallam, A. (1983) *Great Geological Controversies*. Oxford: Oxford University Press.

Hamilton, D. P. (1991) "Research Papers: Who's Uncited Now?" *Science, 251*, 25.

Hildebrand, A. R., and W. V. Boynton (1991) "Cretaceous Ground Zero." *Natural History*, June, 47.

Houghten, R., and G. Woodwell (1989) "Global Climate Change." *Scientific American, 260*, April, 36.

Hoyle, Fred, and Chandra Wickramasinghe (1979) *Diseases from Space*. New York: Harper & Row.

Hsü, K. J. (1972) "When the Mediterranean Dried Up." *Scientific American*, 227, December, 26.

Jacchia, L. G. (1974) "A Meteorite That Missed the Earth." *Sky and Telescope, 48*, 4.

Jeram, Andrew, P. Selden, and Dianne Edwards (1990) "Land Animals in the Silurian: Arachnids and Myriapods from Shropshire, England." *Science, 250*, 658–661.

Jones, P., and T. Wigley (1990) "Global Warming Trends." *Scientific American*, 263, August, 84.

Karl, T. R., R. Heim, and R. Quale (1991) "The Greenhouse Effect in Central North America: If Not Now, When?" *Science, 251*, 1058.

Kerr, Richard A. (1991) "U.S. Bites Greenhouse Bullet and Gags." *Science, 251*, 868.

Khozin, Gregori (1988) *Talking about the Future*, Moscow: Progress Publishers.

Krinov, E. L. (1966) *Giant Meteorites*. Oxford: Pergamon Press.

McAlester, A. Lee (1977) *The History of Life*. Englewood Cliffs, N.J.: Prentice Hall.

Mann, Charles (1991) "Lynn Margulis: Science's Unruly Earth Mother." *Science, 252*, 378.

Martin, Paul S. (1973) "The Discovery of America." *Science, 179*, 969.

Matsui, Takafumi, and Yutaka Abe (1986) "Evolution of an Impact-Induced Atmosphere and Magma Ocean on the Accreting Earth." *Nature, 319*, 303.

Newsom, H. E., and K. W. W. Sims (1991) "Core Formation during Early Accretion of the Earth." *Science*, 252, 926.

Oparin, A. I. (1962) *Life: Its Nature, Origin, and Development*. New York: Academic Press.

Pollard, W. G. (1979) "The Prevalence of Earth-like Planets." *American Scientist*, 67, 653.

Ramskoeld, L., and Hou Xianguang (1991) "New Early Cambrian Animal and Onychophoran Affinities of Enigmatic Metazoans." *Nature, 351*, 225.

Raup, D. M., and J. Sepkoski, Jr. (1986) "Periodic Extinction of Families and Genera." *Science, 231*, 833.

Rhodes, F. H. T., and R. O. Stone (1981) *Language of the Earth*. New York: Pergamon Press.

Ringwood, A. E., T. Kato, W. Hibberson, and N. Ware (1991) "Partitioning of Cr, V, and Mn between Mantles and Cores of Differentiated Planetesimals: Implications for Giant Impact Hypothesis of Lunar Origin." *Icarus*, 89, 122.

Rowland, F. S. (1989) "Chlorofluorocarbons and the Depletion of Stratospheric Ozone." *American Scientist*, 77, 36.

Russell, Dale (1982) "The Mass Extinction of the Late Mesozoic." *Scientific American, 246*, January, 58.

—— (1987) "Models, Paintings, and the Dinosaurs of North America," in *Dinosaurs Past and Present*, eds. Sylvia Czerkas and Everett C. Olson. Seattle: University of Washington Press.

Simkin, Tom, and R. S. Fiske (1983) *Krakatoa 1883*. Washington, D.C.: Smithsonian Institution Press.

Starr, Cecie, and Ralph Taggart (1981) *Biology: The Unity and Diversity of Life*. Belmont, Calif.: Wadsworth Publishing Company.

Stommel, H., and E. Stommel (1983) *Volcano Weather: The Story of the Year Without Summer*. London: Seven Seas Press.

Stringer, C. B. (1990) "The Emergence of Modern Humans." *Scientific American, 263*, December, 98.

Turner, Frederick (1984) "Escape from Modernism." *Harpers, 269*, November, 47.

Turner, Grenville (1988) "Dating of Secondary Events," in *Meteorites and the Early Solar System*, ed. J. Kerridge and M. Matthews. Tucson: University of Arizona Press.

Velikhov, Yevgeni, ed. (1985) *The Night After. . . : Climatic and Biological Consequences of a Nuclear War*. Moscow: Mir Publishers.

Vernadsky, Vladimir (1929) *The Biosphere*. Oracle, Ariz.: Synergetic Press.

Vitaliano, Dorothy B. (1973) *Legends of the Earth: Their Geologic Origins*. Bloomington: Indiana University Press.

White, R. M. (1991) "The Great Climate Debate." *Scientific American, 263*, July, 36.

Williams, G. E. (1989) "Late Precambrian Tidal Rhythmites in South Australia and the History of the Earth's Rotation." *Journ. Geological Soc. London, 146*, January.

INDEX

(Page numbers in italics refer to illustrations.)

B

C

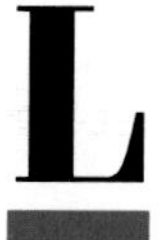

QR

S

ABOUT THE AUTHORS

WILLIAM K. HARTMANN is internationally known as a planetary scientist and also as an astronomical painter and writer. He is especially noted for his research on the evolution of planetary systems and for pioneering the modern theory on how Earth's moon formed. Dr. Hartmann has authored two college textbooks—*Astronomy: The Cosmic Journey* and *Moons and Planets*. In addition to his doctorate in astronomy, he holds an M.S. in geology and has done extensive research on planetary surfaces. He lives in Tuscon, Arizona, where he is a senior editor at the Planetary Science Institute.

RON MILLER is widely known for his astronomical and science-fiction paintings. He was production illustrator for the movies *Dune* and *Total Recall*, and has translated and illustrated two new versions of Jules Verne's *20,000 Leagues Under the Sea* and *Journey to the Center of the Earth*. A former director of the Albert Einstein Spacearium at the National Air & Space Museum in Washington, D.C., Ron Miller lives in Fredericksburg, Virginia.

Other books by William K. Hartmann and Ron Miller are *The Grand Tour*, *Out of the Cradle*, and *Cycles of Fire*. They also edited *In the Stream of Stars: The Soviet/American Space Art Book.*